The Illustrated Flora of Illinois

T0133263

The Illustrated Flora of Illinois
Robert H. Mohlenbrock, *General Editor*

The Illustrated Flora of Illinois

Flowering Plants

Asteraceae, Part 3

Robert H. Mohlenbrock

Southern
Illinois
University Press

Carbondale

Southern Illinois University Press

www.siupress.com

Copyright © 2017 by the Board of Trustees,
Southern Illinois University

All rights reserved

Printed in the United States of America

20 19 18 17   4 3 2 1

Cover and title page illustration: *Eupatoriadelphus fistulosus*
(hollow Joe-pye-weed), by Mark W. Mohlenbrock

Library of Congress Cataloging-in-Publication Data

Names: Mohlenbrock, Robert H., 1931– author.
Title: Flowering plants. Asteraceae, Part 3 / Robert H. Mohlenbrock.
Other titles: Asteraceae | Illustrated flora of Illinois.
Description: Carbondale : Southern Illinois University Press, [2017] | Series:
    The illustrated flora of Illinois | Includes bibliographical references and index.
Identifiers: LCCN 2016056235 | ISBN 9780809336050 (pbk.) | ISBN 9780809336067 (e-book)
Subjects: LCSH: Compositae—Illinois. | Plants—Illinois.
Classification: LCC QK495.C74 M64 | DDC 583/.983—dc23
LC record available at https://lccn.loc.gov/2016056235

To Henry and Alice Barkhausen,

avid conservationists and

my good friends

Contents

Preface

Several volumes in the Illustrated Flora of Illinois series are devoted to the dicoty-
ledonous flowering plants; this volume is the ninth one. It follows publication of
one on ferns, six on monocotyledonous plants, and eight on other dicots.

The concept of the Illustrated Flora of Illinois is to produce a multivolume
flora of the plants of the state of Illinois. For each kind of plant known to occur
in Illinois, complete descriptions and ecological notes are provided, along with a
statement of distribution.

There is no definite sequence for publication of the Illustrated Flora of Illinois.
Volumes will appear as they are completed.

Herbaria from which specimens have been studied are at Eastern Illinois
University, the Field Museum of Natural History, the Gray Herbarium of Harvard
University, the Illinois Natural History Survey, the Illinois State Museum, the Missouri Botanical Garden, the Morton Arboretum, the New York Botanical Garden,
the Shawnee National Forest, Southern Illinois University Carbondale, the United
States National Herbarium, the University of Illinois, and Western Illinois University. In addition, some private collections have been examined. The author is
indebted to the curators and staffs of these herbaria for the courtesies extended.

This is the first volume in the series without illustrations. Lack of sufficient
funding for illustrations along with some physical problems have forced me to
forgo illustrations in this volume. Part 2 of the Asteraceae will be published when
all the illustrations for that part have been completed. My wife, Beverly, assisted
me in several of the herbaria and typed drafts of this manuscript. Madison Preece
prepared the indexes and the glossary. Without the help of all those individuals
and organizations, this book would not have been possible.

The Illustrated Flora of Illinois

Flowering Plants:
Asteraceae, Part 3

County Map of Illinois

Introduction

Flowering plants that form two "seed leaves," or cotyledons, when the seed germinates, are called dicotyledons, or dicots. These far exceed the number of species of monocots, or flowering plants that produce a single "seed leaf" upon germination. This is the ninth volume of the Illustrated Flora of Illinois to be devoted to the dicots of Illinois.

The system of classification adopted for the Illustrated Flora of Illinois was proposed by Thorne in 1968. This system is a marked departure from the more familiar system of Engler and Prantl. This latter system, which is still followed in many regional floras, is out of date and does not reflect the vast information recently gained from the study of cytology, biochemistry, anatomy, and embryology. In fact, the Thorne system no longer depicts many of the relationships adhered to in 1968.

Since the arrangement of orders and families proposed by Thorne is unfamiliar to many, an outline of the orders and families of flowering plants known to occur in Illinois is presented.

Those names in boldface are orders and families that have already been published in this series. The family in capital letters is the one described in this volume of the Illustrated Flora of Illinois.

Order Annonales
Family **Magnoliaceae**
Family **Annonaceae**
Family **Calycanthaceae**
Family **Aristolochiaceae**
Family **Lauraceae**
Family **Saururaceae**

Order Berberidales
Family **Menispermaceae**
Family **Ranunculaceae**
Family **Berberidaceae**
Family **Papaveraceae**

Order Nymphaeales
Family **Nymphaeaceae**
Family **Ceratophyllaceae**

Order Sarraceniales
Family **Sarraceniaceae**

Order Theales
Family **Aquifoliaceae**
Family **Hypericaceae**

Family **Elatinaceae**
Family **Ericaceae**

Order Ebenales
Family **Ebenaceae**
Family **Styracaceae**
Family **Sapotaceae**

Order Primulales
Family **Primulaceae**

Order Cistales
Family **Violaceae**
Family **Cistaceae**
Family **Passifloraceae**
Family **Cucurbitaceae**
Family **Loasaceae**

Order Salicales
Family **Salicaceae**

Order Tamaricales
Family **Tamaricaceae**

Order Capparidales
Family Capparidaceae
Family Resedaceae
Family Brassicaceae

Order Malvales
Family Sterculiaceae
Family Tiliaceae
Family Malvaceae

Order Urticales
Family Ulmaceae
Family Moraceae
Family Urticaceae

Order Rhamnales
Family Rhamnaceae
Family Elaeagnaceae

Order Euphorbiales
Family Thymelaeaceae
Family Euphorbiaceae

Order Solanales
Family Solanaceae
Family Convolvulaceae
Family Cuscutaceae
Family Polemoniaceae

Order Campanulales
Family Campanulaceae

Order Santalales
Family Celastraceae
Family Santalaceae
Family Loranthaceae

Order Oleales
Family Oleaceae

Order Geraniales
Family Linaceae
Family Zygophyllaceae
Family Oxalidaceae
Family Geraniaceae
Family Balsaminaceae
Family Limnanthaceae
Family Polygalaceae

Order Rutales
Family Rutaceae
Family Simaroubaceae
Family Anacardiaceae
Family Sapindaceae
Family Aceraceae
Family Hippocastanaceae
Family Juglandaceae

Order Myricales
Family Myricaceae

Order Chenopodiales
Family Phytolaccaceae
Family Nyctaginaceae
Family Aizoaceae
Family Cactaceae
Family Portulacaceae
Family Chenopodiaceae
Family Amaranthaceae
Family Caryophyllaceae
Family Polygonaceae

Order Hamamelidales
Family Hamamelidaceae
Family Platanaceae

Order Fagales
Family Fagaceae
Family Betulaceae
Family Corylaceae

Order Rosales
Family Rosaceae
Family Mimosaceae
Family Caesalpiniaceae
Family Fabaceae
Family Crassulaceae
Family Penthoraceae
Family Saxifragaceae
Family Droseraceae
Family Staphyleaceae

Order Myrtales
Family Lythraceae
Family Melastomaceae
Family Onagraceae

Order Gentianales
Family Loganiaceae
Family Rubiaceae
Family Apocynaceae
Family Asclepiadaceae
Family Gentianaceae
Family Menyanthaceae

Order Bignoniales
Family Bignoniaceae
Family Martyniaceae
Family Scrophulariaceae
Family Paulowniaceae
Family Plantaginaceae
Family Orobanchaceae
Family Lentibulariaceae
Family Acanthaceae

Order Cornales
Family Vitaceae
Family Nyssaceae

Family Cornaceae
Family Haloragidaceae
Family Hippuridaceae
Family Araliaceae
Family Apiaceae

Order Dipsacales
Family Caprifoliaceae
Family Adoxaceae
Family Valerianaceae
Family Dipsacaceae

Order Lamiales
Family Hydrophyllaceae
Family Heliotropaceae
Family Boraginaceae
Family Verbenaceae
Family Phrymaceae
Family Callitrichaceae
Family Lamiaceae

ORDER ASTERALES
FAMILY ASTERACEAE

Since only one part of one family is included in this book, no general key to the dicot families has been provided. The reader is invited to use my companion book *Guide to the Vascular Flora of Illinois* (2013) for keys to all families of flowering plants in Illinois.

The nomenclature for the species and lesser taxa used in this volume has been arrived at after lengthy study of recent floras and monographs. Synonyms, with complete author citations, that have applied to species and lesser taxa in Illinois are given under each species. A description, while not necessarily intended to be complete, covers the important features of each species.

The common names are the ones used locally in Illinois. The habitat designation is not always the habitat throughout the range of the species, but only for it in Illinois. The overall range for each species is given from the northeastern to the northwestern extremities, south to the southwestern limit, then eastward to the southeastern limit. The ranges have been compiled from various sources, including examination of herbarium material and field studies of my own. A general statement is given concerning the range of each species in Illinois.

Descriptions

This is the third of three volumes describing the family Asteraceae in Illinois. Part 2 will be published when the illustrations are completed.

This family for many years was called the Compositae. Some botanists in the past, with sound reasoning, have divided the family into three families, but I have chosen to follow the current belief that these three segregates should be treated as a single but diverse family.

Worldwide, the family Asteraceae consists of about 23,000 species in about 1,500 genera. In Illinois, I am recognizing 390 species in 120 genera, as well as 21 hybrids and 73 lesser taxa, making it the largest family of flowering plants in the state. Of these 390 species, 130 of them, or 33%, are non-natives. The Poaceae, or grass family, the second largest in Illinois, has 356 species in 102 genera, along with 5 hybrids and 44 lesser taxa.

Worldwide, members of Asteraceae exhibit nearly every growth form, but in Illinois, all species are herbaceous and include annuals, biennials, and perennials. Only one species, *Mikania scandens*, is a vine. All the rest are erect, ascending, or prostrate herbs. Some species have latex. Many plants have taproots or fibrous roots, or both, although there are a number of rhizomatous or stoloniferous species.

With such a large and diverse family, the species exhibit every type of leaf possible. A few species have only basal leaves; others have only cauline leaves; still others have both basal and cauline leaves. In this last group, the basal leaves may or may not be present at flowering time. In those plants with cauline leaves, the leaves may be alternate, opposite, or whorled. Leaves may be simple, unlobed or lobed, or variously compound. Nearly every type of pubescence may be found in the family.

Flowers occur in heads, known as capitulae, with each head having few to numerous flowers. The heads are usually several and are arranged in a variety of inflorescences, called arrays, that may be in the form of corymbs, cymes, panicles, thyrses, or racemes. Occasionally the head may be reduced to one per plant. Each head is surrounded by one or more series of bracts, known as phyllaries. The phyllaries make up the involucre. In a given head, the phyllaries may be equal or unequal. They may be green and herbaceous, or they may have a scarious margin. They may be glabrous or variously pubescent and glandular or eglandular. In *Coreopsis*, *Taraxacum*, and a few other genera, there are tiny bractlets, called calyculi, at the outside base of the phyllaries.

Heads may have ray flowers or disc flowers, or both. Ray flowers have rays that are zygomorphic and often erose or shallowly lobed at the tip. They may have 5 stamens and an inferior ovary. Rays are sometimes referred to as petals by the uninformed. Disc flowers are actinomorphic, usually short-tubular, with 4 or usually 5 lobes. They may have 5 stamens and an inferior ovary.

Experts in this family recognize different types of heads. Radiate heads have peripheral rays that may be pistillate or sterile and central flowers that are disc flowers that are either bisexual or staminate. Liguliferous heads have only ray flowers that are bisexual. Discoid heads have only disc flowers that may be bisexual, only pistillate, or only staminate. In disciform heads, all flowers are disc flowers, but the peripheral flowers have filiform corollas that are usually only pistillate.

All flowers in a head share a common receptacle. The receptacle may be flat or convex. It may bear tiny scales called paleae. The paleae of a receptacle are referred to as chaff. Receptacles without paleae are said to be epaleate or naked. When the paleae are shed, they may leave either a smooth or a pitted receptacle. Occasional hairs, scales, or bristles may also be present on the receptacle.

Although the fruits of the Asteraceae are often referred to as achenes, they are actually cypselae. Achenes are dry, one-seeded fruits that are derived from flowers with a unicarpellate superior ovary. Cypselae are dry, one-seeded fruits that are derived from flowers with a bicarpellate inferior ovary. The cypselae may be crowned or subtended by pappus, which may be in the form of capillary or plumose bristles, awns, or scales. The pappus is usually thought to be the remains of a calyx.

Although molecular phylogenetic studies within the Asteraceae have resulted in different classifications within the family, I am following the more traditional divisions of the family into tribes used in *Flora of North America* (2006).

Below are listed the tribes of Asteraceae in North America, along with the number of genera and species in the United States and in Illinois.

Tribe Mutisieae	7 genera, 14 species in U.S.—None in Illinois
Tribe Cynareae	17 genera, 116 species in U.S.
	10 genera, 34 species, 1 hybrid in Illinois
Tribe Arctotideae	3 genera, 4 species in U.S.—None in Illinois
Tribe Vernonieae	6 genera, 25 species in U.S.
	2 genera, 6 species, 1 hybrid in Illinois
Tribe Cichorieae	50 genera, 230 species in U.S.
	21 genera, 51 species, 5 hybrids in Illinois
Tribe Calenduleae	4 genera, 7 species in U.S.—None in Illinois
Tribe Gnaphalieae	19 genera, 101 species in U.S.
	5 genera, 10 species in Illinois
Tribe Inuleae	3 genera, 5 species in U.S.
	1 genus, 1 species in Illinois
Tribe Senecioneae	29 genera, 167 species in U.S.
	7 genera, 16 species in Illinois
Tribe Plucheae	3 genera, 12 species in U.S.
	1 genus, 2 species in Illinois
Tribe Anthemideae	26 genera, 99 species in U.S.
	9 genera, 24 species in Illinois
Tribe Astereae	77 genera, 719 species in U.S.
	19 genera, 103 species, 1 hybrid in Illinois
Tribe Heliantheae	150 genera, 746 species in U.S.
	42 genera, 140 species, 20 hybrids in Illinois
Tribe Eupatorieae	170 genera, 250 species in U.S.
	8 genera, 25 species, 5 hybrids in Illinois

Because of the large number of taxa of Asteraceae in Illinois, I have divided the tribes into three volumes. Since goldenrods and asters are the largest groups and are often confusing, I elected to treat them, along with other members of the tribe Astereae, in the initial volume. So that each of the three volumes would be more or less comparable in size, I added the tribe Anthemideae to the first book as well.

I do not want to divide any tribe between volumes, so the second volume will comprise the tribes Heliantheae and Senecioneae, although it will treat a few more taxa than the other volumes.

This third book in the series encompasses the other tribes in the Asteraceae: Eupatorieae, Gnaphalieae, Inuleae, Plucheae, Cynareae, Vernonieae, and Cichorieae.

After the description of the Asteraceae and the key to all the genera in Illinois, the description, habitat notes, nomenclatural issues, uses, distribution, and other applicable information are provided for each taxon. Following the name of each taxon are any synonyms that may be pertinent.

Order Asterales

This order consists only of the family Asteraceae.

Family Asteraceae—Aster Family

Annual, biennial, or perennial herbs (in Illinois), rarely a vine (*Mikania scandens*), usually with a taproot, but sometimes rhizomatous or stoloniferous; latex present in some species; leaves basal or cauline, or both, the cauline ones alternate, opposite, or whorled; flowers 2 to numerous in heads, the heads arranged in inflorescences called arrays, sometimes in corymbs, cymes, panicles, racemes, or spikes, less commonly the head solitary; heads subtended by 1 to several bracts, called phyllaries, the phyllaries forming an involucre; heads with ray flowers, disc flowers, or both; ray flowers consisting of a flat lamina of various colors, some of them bisexual, some unisexual, or some neutral, often notched or erose at the tip, sometimes with 5 stamens and an inferior ovary; disc flowers tubular, some of them bisexual, some unisexual, or some neutral, usually 5-lobed (rarely 4-lobed), sometimes with 5 stamens and an inferior ovary; receptacle flat or convex, with or without scales or chaff, known as paleae, that subtend each flower; fruit a cypsela, sometimes crowned by a pappus of capillary bristles, plumose bristles, scales, or awns, or pappus absent.

Key to the Genera of Asteraceae in Illinois

Names appearing in boldface are treated in this book; those followed by I are in the first volume; those followed by II will be in the second volume.

1. Flowering heads with only ray flowers; latex present . . . . .Group 1
1. Flowering heads with disc flowers, the ray flowers present or absent; latex absent.
    2. All the leaves, or at least those on the lower part of the stem, opposite or whorled . . . . .
    . . . . Group 2
    2. None of the leaves opposite or whorled.
        3. Ray flowers present.
            4. Rays yellow or orange. . . . . .Group 3

4. Rays blue, purple, pink, rose, or white, not yellow or orange . . . . Group 4
3. Ray flowers absent.
    5. Leaves simple, entire, toothed, or shallowly lobed . . . . .Group 5
    5. Leaves deeply pinnatifid or pinnately compound . . . . Group 6

Group 1
Flowering heads with only ray flowers; latex present.

1. Flowering heads blue, purple, or pinkish.
    2. Cypselae beakless.
        3. Pappus a crown of scales in 2 to several series.
            4. Pappus of blunt scales; cauline leaves present . . . . .100. **Cichorium**, p. 93
            4. Pappus of aristate scales; cauline leaves absent . . . . 101. **Catananche**, p. 93
        3. Pappus of capillary bristles in 1 series . . . . 109. *Nabalus*, p. 109
    2. Cypselae with a beak about 0.5 mm long up to a long filiform beak.
        5. Pappus of capillary bristles; heads several; peduncles subtended by bractlets.
            6. Beak of cypselae about 0.5 mm long; phyllaries in 1 series. . . . .
            . . . . 107. *Mulgedium*, p. 100
            6. Beak of cypselae usually 1–6 mm long; phyllaries in 2–3 series . . . .
            . . . . 108. *Lactuca*, p. 101
        5. Pappus of plumose bristles; head solitary; peduncles not subtended by bractlets . .
        . . . . 116. **Tragopogon**, p. 126
1. Flowering heads yellow, orange, cream, or whitish.
    7. Flowering heads cream or whitish . . . . 109. *Nabalus*, p. 109
    7. Flowering heads yellow or orange.
        8. Cypselae with a beak 0.5–12.0 mm long.
            9. Cypselae with a stout beak about 0.5 mm long.
                10. Pappus of capillary bristles; heads usually numerous; bractlets at base of
                    peduncle up to 10 . . . . 108. *Lactuca*, p. 101
                10. Pappus of plumose bristles; heads 1 to very few; bractlets at base of
                    peduncle up to 20 . . . . .112. *Leontodon*, p. 122
            9. Cypselae with a filiform beak 1–12 mm long.
                11. Receptacle paleate; pappus of 2 kinds, the outer of capillary bristles, the
                    inner of plumose bristles . . . . .113. *Hypochaeris*, p. 123
                11. Receptacle epaleate; pappus not as above.
                    12. Outer pappus of scales, inner pappus of capillary bristles; flowering
                      heads 1 to few . . . . .120. *Pyrrhopappus*, p. 133
                12. Pappus of all uniform bristles; heads solitary or numerous.
                    13. Pappus of capillary bristles.
                      14. Stems scapose . . . . .103. **Taraxacum**, p. 96
                      14. Stems with some cauline leaves . . . . 104. **Chondrilla**, p. 98
                  13. Pappus of plumose bristles.
                    15. Flowering head solitary, 4–8 cm across. . 116. **Tragopogon**, p. 126
                  15. Flowering heads numerous, up to 2 cm across.
                      16. Peduncles subtended by 5 foliaceous bractlets . . . .
                      . . . . 114. *Helminotheca*, p. 125
                      16. Peduncles subtended by numerous narrow bractlets. . . . .
                      . . . . .115. *Picris*, p. 126

8. Cypselae beakless.
    17. Leaves with spinescent teeth; pappus in 4 series .... 110. *Sonchus*, p. 113
    17. Leaves not spinescent; pappus in 1–2 series, or absent.
       18. Pappus absent.
          19. Flowering head usually solitary; peduncles not subtended by bractlets. . .
          .... 119. *Serinia*, p. 132
          19. Flowering heads few to several; peduncles subtended by 4–5 bractlets . . .
          .... 105. *Lapsana*, p. 98
       18. Pappus present.
          20. Outer pappus of scales, inner pappus of bristles.
             21. Phyllaries up to 35, in 2–5 series; cypselae with 10 ribs.............
             .... 117. *Nothocalais*, p. 129
             21. Phyllaries up to 18, in 1–2 series; cypselae with 10–20 ribs.........
             .... 118. *Krigia*, p. 130
          20. All pappus of bristles.
             22. Pappus of plumose bristles ....112. *Leontodon*, p. 122
             22. Pappus of capillary bristles.
                23. None of the leaves pinnatifid ....111. *Hieracium*, p. 116
                23. Some of the leaves pinnatifid.
                   24. Peduncles subtended by 5–12 bractlets; phyllaries up to 16
                     per head; involucre 4–15 mm across....... 102. *Crepis*, p. 94
                   24. Peduncles subtended by 3–5 bractlets; phyllaries 8 per head;
                     involucre 2–3 mm across .... 106. *Youngia*, p. 99

Group 2

Flowering heads with disc flowers, the ray flowers present or absent; all the leaves, or at least those on the lower part of the stem, opposite or whorled; latex absent.

1. Ray flowers present.
  2. Rays yellow or orange.
    3. Pappus of numerous capillary bristles; plants creeping ..... 51. *Calyptocarpus*, II
    3. Pappus of awns, scales, 1–3 small bristles, or absent; plants upright (procumbent
      in *Sanvitalia*).
      4. Leaves simple, entire, serrate, shallowly lobed, or palmately lobed.
        5. Leaves shallowly palmately lobed; phyllaries in 1 series ...35. *Smallanthus*, II
        5. Leaves not palmately lobed; phyllaries in 2 or more series.
          6. Disc flowers sterile with poorly developed cypselae........43. *Silphium*, II
          6. Disc flowers fertile with well-developed cypselae.
            7. Ray flowers persistent on the cypselae................41. *Heliopsis*, II
            7. Ray flowers deciduous from the cypselae.
              8. Pappus absent or of 1–3 small bristles.
                9. Petioles absent or nearly so; cypselae 4–5 mm long, strongly
                  angular; pappus completely absent........... 37. *Guizotia*, II
                9. Petioles at least 3 mm long; cypselae 1.0–2.5 mm long, weakly
                  angular; pappus absent or of 1–3 small bristles . .50. *Acmella*, II
              8. Pappus of awns or scales.
                10. Cypselae flat.
                  11. Stems winged.
                    12. Cypselae wingless; all leaves opposite... 45. *Verbesina*, II

12. Cypselae winged; some of the leaves alternate . . . .
. . . . .46. *Actinomeris*, II
11. Stems wingless . . . . 47. *Ximenesia*, II
10. Cypselae angled.
13. Cypselae 3-angled; pappus of cypselae of ray flowers with
3 awns . . . . .42. *Sanvitalia*, II
13. Cypselae 2- or 4-angled; pappus of cypselae of ray flowers
with 2 or 4 awns.
14. Cypselae with 2 or 4 stout, barbed awns . . .57. *Bidens*, II
14. Cypselae with 2 small, barbless awns.
15. Rays usually 8 per head; phyllaries in 2 series . . . .
. . . . 54. *Coreopsis*, II
15. Rays of various numbers, not all of them 8 per head;
phyllaries in several series.
16. Cypselae wingless.
17. Ray flowers neutral . . . . 52. *Helianthus*, II
17. Ray flowers fertile . . . . 45. *Verbesina*, II
16. Cypselae winged. . . . . 47. *Ximenesia*, II
4. Leaves deeply pinnately lobed, or 1- to 2-pinnate.
18. Plants aquatic; submerged leaves different from emergent leaves. . . . .
. . . . 58. *Megalodonta*, II
18. Plants terrestrial; submerged leaves absent.
19. Pappus of 20 scales, each with 10 bristles at tip; receptacle bristly . . . . . .
. . . . .59. *Dyssodia*, II
19. Pappus of small scales, not bristle-tipped, with 2 or 4 awns; receptacle
paleate.
20. Phyllaries in 1 series; pappus of unequal scales. . . . . . . 60. *Tagetes*, II
20. Phyllaries in 2 series; pappus of 2–8 awns.
21. Leaves once-divided.
22. Pappus of 2 or 4 stout, barbed awns. . . . . .57. *Bidens*, II
22. Pappus of 2 weak, barbless awns . . . . 54. *Coreopsis*, II
21. Leaves at least twice-divided .56. *Cosmos*, II
2. Rays white, pink, or rose.
23. Leaves pinnately divided or simple and lobed.
24. Some of the leaves pinnately divided; phyllaries in 2 series; pappus present.
25. Leaves and leaflets more than 1 cm wide. . . . . .57. *Bidens*, II
25. Leaflets 1–2 mm wide . . . . .56. *Cosmos*, II
24. Leaves simple and lobed; phyllaries in 2 series; pappus absent . 36. *Polymnia*, II
23. Leaves simple, unlobed.
26. Pappus of ray flowers consisting of 15–20 fimbriate scales . . . 53. *Galinsoga*, II
26. Pappus of ray flowers reduced to a small crown or absent.
27. Outer phyllaries stipitate-glandular. . . . . 45. *Verbesina*, II
27. Outer phyllaries not stipitate-glandular . . . . 49. *Eclipta*, II
1. Ray flowers absent.
28. Flowers unisexual, green.
29. Pistillate involucres nutlike or burlike . . . . 30. *Ambrosia*, II
29. Pistillate involucres not nutlike nor burlike.
30. Heads in racemose spikes, bracteate . . . . .33. *Iva*, II
30. Heads in paniculate spikes, ebracteate. . . . . .34. *Cyclachaena*, II

28. Flowers perfect, not green.
  31. Plants climbing; major phyllaries 4 per head, subtended by short outer phyllaries. . . . . 79. *Mikania*, p. 40
  31. Plants erect; phyllaries not as above.
    32. Some of the leaves divided into 3–7 leaflets or 2- to 3-pinnate.
      33. Leaves once-pinnate . . . . .57. *Bidens*, II
      33. Leaves 2- to 3-pinnate. . . . . .55. *Thelesperma*, II
    32. Leaves simple, sometimes lobed but not compound.
      34. Pappus of 2 or 4 stout, barbed awns . . . . .57. *Bidens*, II
      34. Pappus of capillary bristles, or pappus absent.
        35. Pappus absent; receptacle paleate; stems square . . . . . . 48. *Melanthera*, II
        35. Pappus of capillary bristles; receptacle epaleate; stems not square.
          36. Some or all of the leaves whorled; flowers purple or rose . . . . 73. *Eutrochium*, p. 22
          36. Leaves opposite (if rarely whorled, the flowers white); flowers white, blue, or pink.
            37. Receptacle conical; flowers blue . . . . 74. *Conoclinium*, p. 25
            37. Receptacle flat; flowers white or pink.
              38. Flowers pink . . . . 78. *Fleischmannia*, p. 39
              38. Flowers white.
                39. Phyllaries all of same length . . . . 80. *Ageratina*, p. 41
                39. Phyllaries of different lengths . . . . . . 72. *Eupatorium*, p. 18

Group 3

Plants with both ray and disc flowers; rays yellow or orange; leaves alternate or basal; latex absent.

1. Leaves simple, entire, serrate, or shallowly lobed.
  2. Most or all of the leaves basal.
    3. Head solitary.
      4. Pappus of 5–8 translucent scales . . . . .64. *Tetraneuris*, II
      4. Pappus of capillary bristles . . . . 70. *Tussilago*, II
    3. Heads several . . . . .66. *Packera*, II
  2. Most of the leaves cauline.
    5. Rays reflexed; disc long-columnar or globose.
      6. Disc long-columnar; cauline leaves clasping . . . . .38. *Dracopis*, II
      6. Disc globose; cauline leaves not clasping. . . . . 63. *Helenium*, II
    5. Rays spreading, rarely slightly reflexed; disc globose, conical, or flat.
      7. Pappus entirely of capillary or barbellate bristles.
        8. Leaves spinulose-dentate . . . . 17. *Prionopsis*, I
        8. Leaves not spinulose-dentate.
          9. Upper leaves clasping . . . . 86. *Inula*, p. 54
          9. None of the leaves clasping.
            10. Inflorescence more or less a flat-topped corymb.
              11. Leaves up to 5 (–12) mm wide, some of them usually glandular-punctate . . . . .7. *Euthamia*, I
              11. Leaves usually more than 5 mm wide, not glandular-punctate . . . . .9. *Oligoneuron*, I

10. Inflorescence paniculate, thyrsoid, or in axillary clusters. . 8. *Solidago*, I

7. Pappus of a short crown, awns, or scales, or if capillary bristles present, scales present in addition.

12. Receptacle bristly; pappus of 6–10 awned scales . . . . .62. *Gaillardia*, II

12. Receptacle epaleate or paleate; pappus not as above.

13. Pappus of the cypselae of the disc flowers scaly, with short teeth, a crown, or absent.

14. Outer pappus scaly, inner pappus bristly on the cypselae of the disc flowers.

15. Cypselae of ray flowers thick, of disc flowers flat. . 11. *Heterotheca*, I

15. All cypselae flat . . . . .10. *Chrysopsis*, I

14. Pappus of the cypselae of the disc flowers with short teeth, a crown, or absent.

16. Disc flowers sterile; phyllaries in 2–3 series . . . . . . .43. *Silphium*, II

16. Disc flowers fertile; phyllaries in several series . . . . 39. *Rudbeckia*, II

13. Pappus of the cypselae of the disc flowers with 2–8 awns.

17. Phyllaries gummy. . . . . 16. *Grindelia*, I

17. Phyllaries not gummy.

18. Pappus of 2 awns, these sometimes deciduous.

19. Disc flowers sterile; phyllaries in 2–3 series . . . .43. *Silphium*, II

19. Disc flowers perfect; phyllaries in 1 to several series.

20. Cypselae wingless . . . . .52. *Helianthus*, II

20. Cypselae winged.

21. Stems unwinged . . . . 47. *Ximenesia*, II

21. Stems winged . . . . .46. *Actinomeris*, II

18. Pappus of 5 or more awns, persistent.

22. Disc flowers sterile . . . . 5. *Amphiachyris*, I

22. Disc flowers fertile . . . . .6. *Gutierrezia*, I

1. Leaves simple and deeply divided, or leaves compound.

23. Receptacle epaleate.

24. Rays reflexed; pappus reduced to a short crown or absent; phyllaries in 2 series. . . . . .40. *Ratibida*, II

24. Rays spreading; pappus of barbellate bristles; phyllaries in 1 series.

25. Stems glandular-pubescent; rays 1–3 mm long . . . . .65. *Senecio*, II

25. Stems eglandular; rays 6–10 mm long.

26. Leaves once-pinnatifid . . . . .66. *Packera*, II

26. Leaves 2- to 3-pinnatifid. . . . . .65. *Senecio*, II

23. Receptacle paleate.

27. Receptacle flat; phyllaries in 2–3 series; disc flowers sterile . . . . . .43. *Silphium*, II

27. Receptacle conical; phyllaries in several series; disc flowers fertile.

28. Leaves 2- to 3-pinnate or -pinnatifid . . . . .27. *Cota*, I

28. Leaves lobed or 1-pinnatifid. . . . . .39. *Rudbeckia*, II

Group 4

Plants with both ray flowers and disc flowers; rays not orange nor yellow; latex absent; leaves alternate or basal.

1. Leaves simple and entire, serrate, or shallowly lobed (sometimes with a basal pair of pinnae in *Tanacetum*).

21. Pappus of 2–3 awns; leaves without a basal pair of pinnae.
  22. Stems winged . . . . 45. *Verbesina*, II
  22. Stems unwinged.
    23. Pappus of 2 awns and tiny bristles; plants glabrous; receptacle
      epaleate. . . . . .14. *Boltonia*, I
    23. Pappus of 2–3 awns, without bristles; plants pubescent; receptacle
      paleate . . . . 32. *Parthenium*, II
1. Leaves deeply pinnatifid to 1- to 3-pinnate.
  24. Leaves merely deeply lobed or pinnatisect.
    25. Pappus absent; receptacle epaleate. . . . . 29. *Leucanthemum*, I
    25. Pappus of 2–3 awns; receptacle paleate. . . . . 32. *Parthenium*, II
  24. Leaves 1- to 3-pinnate.
    26. Pappus of capillary bristles; phyllaries more or less squarrose; flowers blue or
      purple . . . . 16. *Machaeranthera*, I
    26. Pappus a low crown, or absent; phyllaries not squarrose; flowers white.
      27. Receptacle paleate, phyllaries in several series; pappus absent.
        28. Plants aromatic; receptacle conical. .25. *Anthemis*, I
        28. Plants not aromatic; receptacle flat . . . . 22. *Achillea*, I
      27. Receptacle epaleate; phyllaries in 2–3 series; pappus a low crown, or absent.
        29. Cypselae 3-ribbed; plants not aromatic . . . . 28. *Tripleurospermum*, I
        29. Cypselae 5-ribbed; plants aromatic . . . . .26. *Matricaria*, I

Group 5

Flowering heads with only disc flowers; leaves alternate or basal and simple,
entire, serrate, shallowly lobed, or pinnatifid (or with a basal pair of pinnae in
Tanacetum); latex absent.

1. Leaves with spine-tipped teeth.
  2. Flowers yellow . . . . .97. ***Centaurea***, p. 75
  2. Flowers purple, pink, pale blue, or white.
    3. Pappus of barbellate bristles; phyllaries in 2 series; stems winged. . . . .
    . . . . 89. ***Onopordum***, p. 58
    3. Pappus of simple bristles, plumose bristles, or a crown of scales; phyllaries in
      several series; stems unwinged.
      4. Pappus a crown of scales; flowers pale blue . . . . 88. ***Echinops***, p. 57
      4. Pappus of bristles; flowers purple, white, or pink.
        5. Pappus of plumose bristles. . . . . .91. ***Cirsium***, p. 61
        5. Pappus of simple bristles . . . . 90. ***Carduus***, p. 59
1. Leaves without spine-tipped teeth.
  6. Outer row of disc flowers appearing ligulate.
    7. Outer phyllaries entire, never spine-tipped.
      8. Phyllaries coriaceous, yellowish, without a hyaline margin . . . .
      . . . . .94. ***Amberboa***, p. 73
      8. Phyllaries thin, green, with a hyaline margin . . . . 93. ***Acroptilon***, p. 72
    7. Outer phyllaries fimbriate or laciniate, some of them spine-tipped.
      9. Involucre 3–4 cm high; pappus 6–12 mm long . . . . . .95. ***Plectocephalus***, p. 74
      9. Involucre 1.0–1.5 cm high; pappus 3 mm long or less. . . . 96. ***Carthamus***, p. 74
  6. Outer row of disc flowers tubular, not appearing ligulate.

10. Flowers greenish.

    11. Phyllaries and fruits with hooked bristles, prickly or appearing prickly.

        12. Flowers unisexual . . . . .31. *Xanthium*, II

        12. Flowers perfect. . . . . .92. **Arctium**, p. 70

    11. Phyllaries and fruits without hooked bristles, not prickly. . . . . 24. *Artemisia*, I

10. Flowers white, purple, cream, yellow, orange, blue, pink, brownish, or rusty.

    13. Phyllaries and fruits with hooked bristles . . . . .92. **Arctium**, p. 70

    13. Phyllaries and fruits without hooked bristles.

        14. Flowers orange; pappus absent or of short scales; some leaves clasping; receptacle paleate . . . . .96. **Carthamus**, p. 74

        14. Flowers not orange; pappus of capillary bristles, a short crown, or 5–8 scales; leaves not clasping; receptacle epaleate.

            15. Flowers yellow, cream, brownish, or rusty.

                16. Phyllaries in 1 series; flowers yellow . . . . 65. *Senecio*, II

                16. Phyllaries in several series; flowers cream, brownish, or rusty.

                    17. Pappus of plumose bristles . . . . 76. **Brickellia**, p. 27

                    17. Pappus of capillary bristles.

                        18. Heads leafy bracted . . . . .84. **Gnaphalium**, p. 52

                        18. Heads not leafy bracted. . . . ,. . 82. **Pseudognaphalium**, p. 49

            15. Flowers pink, purple, or white.

                19. Flowers pink or purple.

                    20. Flowers in glomerules, subtended by a 3-lobed bract . . . . .98. *Elephantopus*, p. 85

                    20. Flowers and bracts not as above.

                        21. Phyllaries in 1 series, subtended by calyxlike bracts; leaves basal. . . . . .71. *Petasites*, II

                        21. Phyllaries in several series; leaves not basal.

                          22. Pappus of plumose or barbellate bristles. . . 77. **Liatris**, p. 29

                          22. Pappus of simple bristles.

                              23. Pappus in a double series . . . . .99. **Vernonia**, p. 86

                              23. Pappus in a single series.

                                  24. Capillary bristles united at base. . 85. **Gamochaeta**, p. 53

                                24. Capillary bristles not united at base . . 87. **Pluchea**, p. 55

            19. Flowers white.

                25. Phyllaries in 1 series.

                    26. Calyxlike bracts at base of phyllaries; leaves basal . . . . .71. *Petasites*, II

                    26. Calyxlike bracts absent; leaves cauline.

                      27. Phyllaries 5; flowers 5 per head . . . . 69. *Arnoglossum*, II

                      27. Phyllaries more than 5; flowers more than 5 per head . . . . .67. *Erechtites*, II

            25. Phyllaries in 2 to several series.

                28. Some of the leaves hastate . . . . .68. *Hasteola*, II

                28. None of the leaves hastate.

                    29. Pappus a crown of short scales . . . . .21. *Tanacetum*, I

                    29. Pappus of capillary bristles.

                      30. Phyllaries in 2 series, not scarious . . . .20. *Brachyactis*, I

                      30. Phyllaries in several series, scarious.

31. Most of the leaves near the base of the plant . . . .
. . . . 81. *Antennaria*, p. 43
31. Most of the leaves cauline . . . . . . 83. *Anaphalis*, p. 51

Group 6

Flowering heads with only disc flowers; leaves alternate or basal, pinnatifid to pinnately compound; latex absent.

1. Flowers greenish.
  2. Pistillate involucre with 2 flowers and 2 beaks; receptacle paleate . . . 30. *Ambrosia*, II
  2. Involucres not as above; receptacle epaleate or woolly . . . . 24. *Artemisia*, I
1. Flowers yellow or greenish yellow.
  3. Pappus of capillary bristles; phyllaries in 1 series . . . . 65. *Senecio*, II
  3. Pappus of 12–20 hyaline scales, of short awns, or absent; phyllaries in 2–4 series.
    4. Pappus of 12–20 hyaline scales. . . . . 61. *Hymenopappus*, II
    4. Pappus a crown of short scales, or absent.
      5. Receptacle conic; plants aromatic. . . . . 26. *Matricaria*, I
      5. Receptacle flat; plants not aromatic . . . . 21. *Tanacetum*, I

Tribe Eupatorieae Cass.

Herbs (in Illinois) or shrubs; leaves cauline, usually opposite or whorled, simple or sometimes deeply dissected; heads discoid in corymbs, panicles, or spikes, not subtended by bractlets; involucre hemispheric, conical, obconic, or campanulate; phyllaries in 2–8 series, unequal to equal, sometimes with scarious margins; receptacle flat or convex or conical, mostly epaleate; ray flowers absent; disc flowers white, blue, pink, or purple, bisexual, fertile, the corolla usually 5-lobed; cypselae various, angled and/or ribbed, glabrous or pubescent; pappus of barbellate to plumose bristles or scales or awns or both bristles and scales.

There are 170 genera in this tribe and nearly 2,500 species. Nine genera occur in Illinois.

Arrangement of the leaves, color of the flowers, and characters of the phyllaries and pappus are the major considerations in the separation of genera within the tribe.

The sequence of genera follows that in *Flora of North America*.

Key to the Genera of Eupatorieae in Illinois

1. Vines; phyllaries 4 per flowering head. . . . . 79. *Mikania*
1. Erect or ascending herbs, not viny; phyllaries more than 4 per flowering head.
  2. Most of the leaves alternate.
    3. Flowers purple, borne in spikes . . . . 77. *Liatris*
    3. Flowers white or cream, borne in corymbs . . . . .76. *Brickellia*
  2. Leaves opposite or whorled.
    4. Leaves whorled; flowers pink or rose . . . . .73. *Eutrochium*
    4. Leaves opposite or, if whorled, the flowers white, otherwise blue or pink.
      5. Receptacle conical; flowering heads blue.
        6. Leaves 3-nerved from the base; involucre hemispheric; pappus of about 30 barbellate bristles . . . . 74. *Conoclinium*
        6. Leaves 1-nerved; involucre campanulate; pappus of 5–6 aristate scales . . . . .
. . . . 75. *Ageratum*
      5. Receptacle flat or convex; flowering heads white or pink.

7. Flowering heads pink .... 78. *Fleischmannia*
7. Flowering heads white.
   8. Phyllaries all the same length .... 80. *Ageratina*
   8. Phyllaries of different lengths .... 72. *Eupatorium*

72. **Eupatorium** L.—Boneset; Thoroughwort

Perennial herbs; stems erect; leaves cauline, opposite, rarely alternate or whorled, simple or deeply pinnately divided, usually glandular-dotted; heads discoid, usually borne in corymbs, not subtended by bractlets; involucre campanulate to obconic to ellipsoid, up to 3 (–5) mm in diameter; phyllaries up to 15 in 2–3 series, usually 2- or 3-nerved, mostly unequal, usually with scarious margins, pubescent; receptacle flat or convex, epaleate; disc flowers tubular, 3–5 (–15) per head, white, bisexual, perfect, the corolla lobes 5; cypselae prismatic, 5-ribbed, usually glabrous, glandular-dotted; pappus of up to 50 barbellate bristles in 1 series.

As considered here, *Eupatorium* consists of 40–45 species in eastern North America, Europe, and Asia. Six species and 1 hybrid are known from Illinois.
1. Leaves deeply pinnately divided into filiform segments .... 2. *E. capillifolium*
1. Leaves simple, not divided into filiform segments.
   2. Leaves with distinct petioles .... 6. *E. serotinum*
   2. Leaves sessile, subsessile, or perfoliate.
      3. Leaves, or most of them, perfoliate .... 4. *E. perfoliatum*
      3. Leaves sessile or subsessile.
         4. Leaves rounded or subcordate at base .... 7. *E. sessilifolium*
         4. Leaves tapering to base.
            5. Some of the leaves whorled, narrowly lanceolate, usually more than six times longer than wide .... 3. *E. hyssopifolium*
            5. Leaves not whorled, lanceolate to elliptic, less than six times longer than wide.
               6. Leaves serrate throughout .... 4. *E. perfoliatum*
               6. Leaves serrate only in the upper two-thirds of the leaf.
                  7. Leaves 3-veined, not rugose .... 1. *E. altissimum*
                  7. Leaves several-nerved, rugose .... 5. *E.* X *polyneuron*

1. **Eupatorium altissimum** L. Sp. Pl. 2:837. 1753.

Perennial herbs from rhizomes; stems erect, branched above, to 2 m tall, densely short-pubescent; leaves opposite, simple, lanceolate, acuminate at the apex, tapering to the sessile base, to 12 cm long, to 2 cm wide, 3-nerved from the base, serrate only in the upper two-thirds, densely puberulent, glandular-dotted; heads numerous in corymbs, discoid; involucre campanulate; phyllaries up to 10, in 2–3 series, oblong, obtuse at the apex, 1–4 mm long, densely puberulent; receptacle flat, epaleate; ray flowers absent; disc flowers 5, tubular, bisexual, fertile, white, the corolla 5-lobed, 3.0–3.5 mm long; cypselae prismatic, 5-ribbed, pubescent, 2–3 mm long; pappus of up to 40 barbellate bristles 3.5–4.0 mm long.

Common Name: Tall boneset; tall thoroughwort.
Habitat: Woods, fields, pastures, prairies, disturbed soil.
Range: Massachusetts to Minnesota, south to Texas and Florida.
Illinois Distribution: Common throughout Illinois.

This species is distinguished by its sessile, lanceolate leaves that are serrate only in the upper two-thirds of the margins. It is common in a variety of dry habitats.

Eupatorium altissimum flowers from August to October.

2. **Eupatorium capillifolium** (Lam.) Small, Mem. Torrey Club 5:311. 1894.
Artemisia capillifolia Lam. Encycl. 1:267. 1783.

Annual or perennial herbs from a thickened caudex; stems erect, much branched, densely puberulent to hirsute, to 3 m tall; leaves mostly alternate, deeply pinnately divided, to 10 cm long, the segments filiform, to 1 mm wide, sessile, glabrous or nearly so, glandular-dotted; heads numerous in dense panicles, discoid; involucre campanulate; phyllaries up to 10, in 2–3 series, linear, acuminate and mucronate at the apex, glabrous or nearly so, to 2.5 mm long; receptacle flat, epaleate; ray flowers absent; disc flowers usually 5, tubular, bisexual, fertile, white, the corolla 5-lobed, 2.0–2.5 mm long; cypselae prismatic, 5-ribbed, puberulent, 1.0–1.5 mm long; pappus of 20–30 barbellate bristles 2.0–2.5 mm long.

Common Name: Dog fennel.
Habitat: Old field (in Illinois).
Range: Pennsylvania to Missouri, south to Texas and Florida.
Illinois Distribution: Known only from Alexander County where several plants were growing in an old field.

This species is recognized by its large panicles of small whitish flowering heads, its deeply pinnately divided leaves with filiform segments, and its tall stature.

Several plants were found in an old field adjacent to Interstate Highway 57 in Alexander County, about one mile north of the bridge over the Mississippi River, on September 14, 2008.

Eupatorium capillifolium flowers during August and September.

3. **Eupatorium hyssopifolium** L. Sp. Pl. 2:836. 1753.
Perennial herbs from short rhizomes; stems erect, branched, to 1.5 m tall, densely pubescent; leaves simple, opposite or whorled, on the uppermost alternate, linear-lanceolate to lanceolate, more than six times longer than wide, acute to obtuse at the apex, tapering to the sessile base, to 5 cm long, to 5 mm wide, 3-nerved from the base, entire to serrulate, scabrous, with tufts of axillary leaves; heads numerous in a more or less flat-topped corymb, discoid; involucre campanulate; phyllaries 8–10 in 2–3 series, linear to narrowly oblong, obtuse at the apex, densely puberulent, to 5 mm long; receptacle flat, epaleate; ray flowers absent; disc flowers 5, tubular, bisexual, fertile, white, the corolla 5-lobed, 3.0–3.5 mm long; cypselae prismatic, 5-ribbed, pubescent, 2–3 mm long; pappus of 20–30 barbellate bristles 3.5–4.0 mm long.

Common Name: Hyssop-leaved boneset.
Habitat: Dry open woods.
Range: Massachusetts to Wisconsin, south to Texas and Florida.
Illinois Distribution: Known only from Pope County.

This is the only *Eupatorium* in Illinois with whorled leaves that are 1 to 5 mm wide. Its original location was in a cemetery in Illinois where it was discovered by John Schwegman.

Eupatorium hyssopifolium flowers from August to October.

4. **Eupatorium perfoliatum** L. Sp. Pl. 2:838. 1753.
Eupatorium truncatum Muhl. ex Willd. Sp. Pl. 3:1751–52. 1803.
Eupatorium perfoliatum L. f. *purpureum* Britt. Bull. Torrey Club 17:124. 1890.
Eupatorium perfoliatum L. f. *trifolium* Fassett, Rhodora 27:55–56. 1925.
Eupatorium perfoliatum L. f. *truncatum* (Muhl. ex Willd.) Fassett, Rhodora 27:55. 1925.

Perennial herbs from a woody caudex; stems erect, branched, to 1.5 m tall, densely villous; leaves simple, opposite, rarely whorled, lanceolate, acute to acuminate at the apex, rounded or truncate at the connate base, to 15 cm long, to 4.5 cm wide, several-nerved, serrate to crenate, pilose or villous, strongly rugose, glandular-dotted; heads numerous in a somewhat flat-topped corymb, discoid; involucre campanulate; phyllaries up to 10, in 1–2 series, oblong to lanceolate, acute at the apex, pubescent, to 4.5 mm long, glandular-dotted; receptacle flat, epaleate; ray flowers absent; disc flowers 7–10, tubular, bisexual, fertile, white or very rarely purple, the corolla 5-lobed, 2.5–3.0 mm long; cypselae prismatic, 5-ribbed, densely pubescent, 1.5–2.0 mm long; pappus of 20–30 barbellate bristles 3.0–3.5 mm long.

Common Name: Perfoliate boneset; common boneset.
Habitat: Marshes, bogs, calcareous fens, moist sand flats, wet prairies, other wet areas.
Range: Nova Scotia to Manitoba, south to Texas and Florida.
Illinois Distribution: Common throughout the state.

Plants with opposite leaves that are connate at the base and with white flowers are the typical form. Plants with whorled leaves and rounded connate leaf bases and white flowers may be known as f. *trifolium*. Plants with opposite truncate-connate leaf bases and white flowers may be known as f. *truncatum*. These plants were originally described as a distinct species; some botanists believe it to be a hybrid between *E. perfoliatum* and *E. serotinum*. Plants with opposite rounded connate leaf bases and purple flowers may be known as f. *purpureum*. All these forms have been found sparingly in Illinois.

Eupatorium perfoliatum flowers from July to October.

5. **Eupatorium X polyneuron** (F. J. Herrm.) Wunderlin, Ann. Mo. Bot. Gard. 59:472. 1972.
Eupatorium cuneatum Engelm. ex Torr. & Gray, N. Am. Fl. 2:88. 1841.
Eupatorium perfoliatum L. var. *cuneatum* (Engelm. ex Torr. & Gray) Engelm. ex Gray, Fl. N. Am. 1:100. 1884.
Eupatorium serotinum Michx. var. *polyneuron* F. J. Herm. Rhodora 40:86. 1938.

Perennial herbs from a woody caudex; stems erect, branched, to 1.5 m tall, more or less hirtellous; leaves opposite, simple, narrowly lanceolate, acuminate at the apex, tapering to the sessile base, to 10 cm long, to 1.5 cm wide, hirtellous, somewhat rugose, many-nerved, crenate to serrate in the upper two-thirds of the margins; heads numerous in corymbs, discoid; involucre campanulate; phyllaries 7–12 in 1–2 series, narrowly oblong, acute at the apex, puberulent, glandular-dotted; receptacle flat, epaleate; ray flowers absent; disc flowers 7–14, tubular, white, the corolla 5-lobed, 2.5–3.0 mm long; cypselae prismatic, ribbed, 1.0–1.7 mm long, pubescent; pappus of 20–30 barbellate bristles 2.0–3.5 mm long.

Common Name: Hybrid boneset.
Habitat: Wet ground.
Range: Ohio to Missouri, south to Arkansas and Louisiana.
Illinois Distribution: Scattered in Illinois, but not common.

This is considered to be the hybrid between *E. perfoliatum* L. and *E. serotinum* Michx. It sometimes has been called *E. perfoliatum* var. *cuneatum*.
   I previously erroneously referred to this plant as *E. X platyneuron*.
   This hybrid flowers from July to October.

6. **Eupatorium serotinum** Michx. Fl. Bor. Am. 2:100. 1803.
   Perennial herbs with a short caudex; stems erect, branched, often purplish, to 2 m tall, pubescent; leaves simple, opposite, lanceolate to ovate, acuminate at the apex, tapering to the petiolate base, to 10 cm long, to 6 cm wide, 3- or 5-nerved from the base, sharply serrate, usually pubescent, glandular-dotted, the petioles 10 mm long or longer; heads numerous in a corymb, discoid; involucre campanulate; phyllaries 8–12 in 1–3 series, linear-oblong, obtuse at the apex, densely pubescent, glandular-dotted; receptacle flat, epaleate; ray flowers absent; disc flowers 10–15, tubular, bisexual, fertile, white, the corolla 5-lobed, 2.5–3.0 mm long; cypselae prismatic, ribbed, pubescent, 1.0–1.5 mm long; pappus of 20–30 barbellate bristles 2.0–2.5 mm long.

Common Name: Late boneset; late thoroughwort.
Habitat: Moist open woods, pastures, disturbed areas.
Range: Massachusetts to Minnesota, south to Nebraska, Texas, and Florida.
Illinois Distribution: Common throughout the state.

This is the only species of *Eupatorium* in Illinois that has petioles at least 10 mm long. The similar-appearing, petiolate *Ageratina altissima* has green stems, whereas the stems of *E. serotinum* are usually purple.
   Eupatorium serotinum flowers from July to October.

7. **Eupatorium sessilifolium** L. Sp. Pl. 2:837. 1753.
   Perennial herbs from short rhizomes; stems erect, branched or unbranched, up to 2 m tall, glabrous or sometimes pubescent; leaves simple, opposite or the uppermost alternate, lanceolate to lance-ovate, acuminate at the apex, rounded or subcordate at the sessile base, to 18 cm long, to 6 cm wide, thin or firm, several-nerved, serrate, glabrous or nearly so, glandular-dotted; heads numerous in

corymbs, discoid; involucre campanulate; phyllaries up to 15 in 2–3 series, linear to narrowly oblong, obtuse to acute at the apex, villous, glandular-dotted, to 1.5 mm long; receptacle flat, epaleate; ray flowers absent; disc flowers 5, tubular, bisexual, fertile, white, the corolla 5-lobed, 3.0–3.5 mm long; cypselae prismatic, ribbed, pubescent, 2–3 mm long; pappus of 30–40 barbellate bristles 3–4 mm long.

Two varieties occur in Illinois.
a. Leaves lanceolate, 2–4 cm wide, thin; branches of the inflorescence glabrous . . . .
. . . . .7a. *E. sessilifolium* var. *sessilifolium*
a. Leaves lance-ovate, to 6 cm wide, firm; branches of the inflorescence pubescent . . . . . .
. . . . .7b. *E. sessilifolium* var. *brittonianum*

7a. **Eupatorium sessilifolium** L. var. **sessilifolium**
Leaves lanceolate. 2–4 cm wide; branches of the inflorescence glabrous.

Common Name: Upland boneset.
Habitat: Woods.
Range: Massachusetts to Minnesota, south to Arkansas and Georgia; Kansas.
Illinois Distribution: Occasional in the southern half of Illinois, rare or absent elsewhere.

This appears to be the more common of the 2 varieties in Illinois.
This plant flowers from August to October.

7b. **Eupatorium sessilifolium** L. var. **brittonianum** Porter, Bull. Torrey Club 19:129. 1892.
Leaves lance-ovate, up to 6 cm wide; branches of the inflorescence pubescent.

Common Name: Sessile-leaved boneset.
Habitat: Dry woods, prairies, dry or mesic savannas.
Range: New York to Minnesota, south to Missouri and North Carolina.
Illinois Distribution: Scattered in the southern half of the state.

This variety has firmer, wider leaves that var. *sessilifolium*.
Eupatorium sessilifolium var. *brittonianum* flowers from August to October. Most of our specimens were found in dry woods, although one plant occurred for several years in the DeSoto Railroad Prairie in Jackson County. In the northeastern counties, it occurs in savannas.

73. **Eutrochium** Raf.—Joe-pye-weed
Robust perennial herbs; stems erect, usually unbranched, sometimes purple, sometimes glaucous; leaves simple, whorled, serrate, glandular-dotted; heads numerous in flat-topped or round-topped corymbs, discoid, not subtended by bractlets; involucre cylindric; phyllaries in 5–6 series, usually pink or purplish, appressed, unequal, glabrous or pubescent, glandular-dotted; receptacle flat or convex, epaleate; ray flowers absent; disc flowers up to 20 (–30) per head, pink or purplish, the corolla 5-lobed; cypselae prismatic, 5-ribbed, glandular-dotted; pappus of 25–40 pink, purplish, or cream barbellate bristles in 1 series.

This genus consists of 5 species, all in North America.

All of our species were originally placed in *Eupatorium* by Linnaeus. King and Robinson removed these species from *Eupatorium* in 1970, placing them in *Eupatoriadelphus*. However, Rafinesque had already removed them from *Eupatorium* in 1838, placing them in *Eutrochium*, which is the correct generic name.

Eutrochium is readily distinguished by its tall stature, whorled leaves, and pink or purplish disc flowers.

Three species occur in Illinois.

1. Heads with 5–7 flowers; stems usually glaucous, at least when young; inflorescence round-topped.
   2. Stems hollow; leaves often more than 4 in a whorl . . . . .3. *E. fistulosum*
   2. Stems pithy; leaves usually 4 in a whorl. . . . . 2. *E. purpureum*
1. Heads with 9–22 flowers; stems not glaucous; inflorescence more or less flat-topped . . .
   . . . . 1. *E. maculatum*

1. **Eutrochium maculatum** (L.) E. E. Lamont, Sida 21:902. 2004.
Eupatorium maculatum L. Cent. Pl. I, 27. 1755.
Eupatorium purpureum L. var. *maculatum* (L.) Darl. Fl. Cestr. 453. 1837.
Eupatorium bruneri Gray, Syn. Fl. N. Am. 1:96. 1884.
Eupatorium maculatum L. var. *bruneri* (Gray) Breitung, Can. Field-Nat. 61:98. 1947.
Eupatorium maculatum L. f. *faxonii* Fern. Rhodora 47:195. 1945.
Eupatoriadelphus maculatus (L.) R. M. King & H. Robins. Phytologia 19:432. 1970.

Perennial herbs from a thickened caudex; stems erect, most unbranched except near the top, solid, to 2 m tall, purple-speckled, glabrous to sparsely pubescent, often viscid; leaves simple, in whorls of (3–) 4–5 (–6), narrowly lanceolate to lance-ovate, acute to acuminate at the apex, tapering to the petiolate base, to 20 cm long, to 7.5 (–9.0) cm wide, glabrous or sparsely pubescent, glandular-dotted; heads numerous in flat-topped corymbs, discoid; involucre cylindric; phyllaries in 5–6 series, unequal, oblong, obtuse at the apex, mostly glabrous, to 6 mm long; receptacle flat to convex, epaleate; ray flowers absent; disc flowers 9–22, tubular, bisexual, perfect, purple to rose-purple, rarely white, the corolla 5-lobed, 5–7 mm long; cypselae prismatic, 5-ribbed, pubescent, 3.0–4.5 mm long; pappus of about 40 pink or purplish barbellate bristles 2.0–4.5 mm long.

Common Name: Spotted Joe-pye-weed.
Habitat: Calcareous fens, marshes, sedge meadows.
Range: Newfoundland to Minnesota, south to Iowa, Illinois, and North Carolina.
Illinois Distribution: Common in northern Illinois, rare in southern Illinois.

This species has leaves mostly in whorls of 4 or 5. Its stems are usually purple-speckled, although some plants in Illinois have completely purple stems. Plants with puberulent stems throughout have been called var. *bruneri*. The flowers are more rose-purple than in the other 2 species in Illinois. This species is very showy when in flower. Rare white-flowered specimens are seen. They are called f. *faxonii*. This species is often an indicator of fens.

Eutrochium maculatum flowers from June to October.

2. **Eutrochium purpureum** (L.) E. E. Lamont, Sida 21:902. 2004.
Eupatorium purpureum L. Sp. Pl. 2:838. 1753.
Eupatorium falcatum Michx. Fl. Bor. Am. 2:99. 1803.
Eupatoriadephus purpureus (L.) R. M. King & H. Robins. Phytologia 19:432. 1970.

Robust perennial herbs from a woody caudex; stems erect, branched only near the top, green but usually purple at the nodes, not speckled with purple, to 3 m tall, glabrous or sparsely pubescent, glaucous when young; leaves simple, in whorls of (3–) 4, ovate-lanceolate to ovate, acute to acuminate at the apex, tapering to the petiolate base, to 25 cm long, to 15 cm wide, serrate, glabrous or pubescent, at least beneath, glandular-dotted, several-nerved, the petioles 5–20 mm long; heads numerous in a round-topped corymb, discoid; involucre cylindric; phyllaries in 5–6 series, unequal, oblong, obtuse at the apex, pink, mostly glabrous, to 6 mm long; receptacle convex, epaleate; ray flowers absent; disc flowers 5–7, tubular, bisexual, perfect, pink or purplish, the corolla 5-lobed, 5.0–6.5 mm long; cypselae prismatic, 5-ribbed, pubescent, 3.0–4.5 mm long; pappus of up to 40 pinkish barbellate bristles 2.0–4.5 mm long.
Two varieties occur in Illinois.
a. Lower surface of leaves glabrous or sparsely pubescent ....
....2a. *E. purpureum* var. *purpureum*
a. Lower surface of leaves densely puberulent...........2b. *E. purpureum* var. *holzingeri*

2a. **Eutrochium purpureum** (L.) E. E. Lamont var. **purpureum**
Lower surface of leaves glabrous or sparsely pubescent.

Common Name: Purple Joe-pye-weed.
Habitat: Woods, savannas.
Range: Massachusetts to Minnesota, south to Nebraska, Oklahoma, Louisiana, and Florida.
Illinois Distribution: Occasional in northern Illinois.

The glabrous or sparsely pubescent leaves distinguish this variety from var. *holzingeri*. The green stems with usually purple nodes are distinctive.
This variety flowers from June to October.

2b. **Eutrochium purpureum** (L.) E. E. Lamont var. **holzingeri** (Rydb.) E. E. Lamont, Sida 21:902. 2004.
Eupatorium holzingeri Rydb. Brittonia 1:97. 1931.
Eupatorium purpureum L. var. *holzingeri* (Rydb.) E. E. Lamont, Phytologia 69:468. 1990.

Lower surface of leaves densely puberulent.

Common Name: Hairy purple Joe-pye-weed.
Habitat: Woods, savannas.
Range: Wisconsin to Minnesota, south to Kansas to Arkansas.
Illinois Distribution: Occasional in northern Illinois.

This variety flowers from June to October.

3. **Eutrochium fistulosum** (Barratt) E. E. Lamont, Sida 21:904. 2004.
Eupatorium fistulosum Barratt, Eupatoria Verticillata no. 1. 1841.
Eupatoriadelphus fistulosus (Barratt) R. M. King & H. Robins. Phytologia 19:432.
  1990.

Robust perennial herbs from a woody caudex; stems erect, branched only near
the top, usually purple, strongly glaucous, hollow, usually glabrous, to 3 m tall;
leaves simple, in whorls of (4–) 5–6, elliptic-lanceolate, acute to acuminate at the
apex, tapering to the petiolate base, to 25 cm long, to 8 cm wide, crenate, glabrous
or sparsely pubescent, sparsely glandular-dotted, several nerved, the petiole 10–30
mm long; heads numerous in a round-topped corymb, discoid; involucre cylindric;
phyllaries in 5–6 series, unequal, oblong, obtuse at the apex, pink, more or less glabrous, to 6 mm long; receptacle convex, epaleate; ray flowers absent; disc flowers
5–7, tubular, bisexual, perfect, pink or purplish, the corolla 5-lobed, 4.5–8.0 mm
long; cypselae prismatic, 5-ribbed, usually puberulent, 3.0–4.5 mm long; pappus of
up to 40 pinkish barbellate bristles 3–4 mm long.

Common Name: Hollow Joe-pye-weed; trumpet-weed.
Habitat: Low, wet ground, marshes, swamp forests.
Range: Maine to Wisconsin, south to Texas and Florida.
Illinois Distribution: Occasional in the southern half of Illinois, less common or rare
  elsewhere.

This species is recognized by its hollow, strongly glaucous stems and its leaves
usually in whorls of 5 or 6. It occupies wetter habitats than the similar-appearing
E. purpureum.

The common name trumpet-weed comes from the fact that pioneer children
used these plants to make musical sounds. They would cut sections of the hollow stems and puncture holes in them. Then, by blowing into the end of a section while covering one or more of the puncture holes, they could make musical
sounds.

Eutrochium fistulosum flowers from July to September.

74. **Conoclinium** DC.—Mistflower

Perennial herbs from slender rhizomes; stems erect, scarcely branched; leaves
simple, opposite, glabrous or pubescent, gland-dotted; heads numerous, in corymbs, discoid, without bractlets at base; phyllaries up to 25, in 2–3 series, more
or less equal; receptacle conical, epaleate; ray flowers absent; disc flowers up to
75 per head, tubular, blue or purple, bisexual, fertile, the corolla 5-lobed; cypselae
prismatic, 5-ribbed, glabrous or pubescent; pappus of 25–30 barbellate bristles in
one series.

This genus comprises 4 species, all of them in the United States or Mexico.

The single species found in Illinois has a striking resemblance to the ornamental *Ageratum*. *Conoclinium* has cypselae with a pappus of 25–30 barbellate bristles.
Ageratum has cypselae with a pappus of 5–6 aristate scales.

1. **Conoclinium coelestinum** (L.) DC. Prodr. 5:135. 1836.

Eupatorium coelestinum L. Sp. Pl. 2:838. 1753.

Eupatorium coelestinum L. f. *illinoense* Benke, Rhodora 45:36. 1943.

Perennial herbs from slender rhizomes and stolons; stems erect, branched, pubescent, to 75 cm tall; leaves simple, opposite, deltate to ovate, to 10 cm long, to 5 cm wide, acute at the apex, truncate at the petiolate base, crenate or serrate, pubescent, 3- or 5-nerved from the base, sometimes with axillary clusters, the petioles 10–20 mm long; heads numerous in rather compact corymbs, discoid; involucre hemispheric; phyllaries up to 20 (–25), in 2 or 3 series, linear to narrowly lanceolate, acuminate, usually glabrous, equal; receptacle conical, epaleate; ray flowers absent; disc flowers 35–75, tubular, blue, rarely purple or white, bisexual, fertile, the corolla 5-lobed, 1.5–2.5 mm long; cypselae prismatic, 5-ribbed, glabrous, 1.0–1.5 mm long; pappus of about 30 barbellate bristles 5–8 mm long.

Common Name: Mistflower.

Habitat: Moist ground.

Range: New York to Ontario, south to Nebraska, Texas, and Florida.

Illinois Distribution: Common in the southern two-thirds of Illinois; rare northward
    where it may have been introduced.

This attractive species often covers large areas of wetlands in the southern two-thirds of Illinois, the blue flowering heads giving it a misty appearance from a distance. A white-flowered form has been found near Marion in Williamson County in 1931 and was named *Eupatorium coelestinum* f. *illinoense*. This varietal epithet has not been transferred to *Conoclinium*.

Although originally placed in *Eupatorium*, *Conoclinium* differs in its blue flowering heads and conical receptacle. It has a striking resemblance to the garden ornamental *Ageratum conyzoides*, but this latter species has a pappus of 5–6 aristate scales rather than 25–30 barbellate bristles.

Conoclinium coelestinum flowers from July to October.

75. **Ageratum** L.—Ageratum

Annual or perennial herbs; stems erect (in Illinois) or decumbent; leaves simple, opposite, one-nerved, glabrous or pubescent, glandular-dotted; heads numerous, in dense corymbs, discoid; involucre campanulate; phyllaries up to 35, in 2–3 series, 2-nerved, more or less equal; receptacle conical, epaleate; ray flowers absent; disc flowers numerous, tubular, blue or sometimes white, bisexual, fertile, the corolla 5-lobed; cypselae prismatic, 4- or 5-ribbed, glabrous or pubescent; pappus of 5 or 6 aristate scales.

This genus comprises about 40 species, all in the United States, Mexico, and Central America.

Only 1 species is found in Illinois.

1. **Ageratum conyzoides** L. Sp. Pl. 2:839. 1753.

Annual or perennial herbs; stems erect, usually branched, villous, to 1 m tall; leaves simple, opposite, oblong to ovate, to 8 cm long, to 5 cm wide, acute at the apex, truncate or rounded at the petiolate base, serrate, several-nerved, sparsely pubescent, glandular-ciliate; heads numerous, in dense corymbs, discoid; involucre campanulate; phyllaries up to 35, in 2–3 series, oblong-lanceolate, glabrous or puberulent, 2-nerved, to 4 mm long, more or less equal; receptacle conical, epaleate; ray flowers absent; disc flowers numerous, tubular, blue, bisexual, fertile, the corolla 5-lobed, to 4.5 mm long; cypselae prismatic, 4- or 5-ribbed, sparsely pubescent, 1.5–3.0 mm long; pappus of 5–6 aristate scales up to 3 mm long.

Common Name: Ageratum.
Habitat: Disturbed soil is a vacant lot.
Range: Native to South America; occasionally escaped from cultivation in the United States.
Illinois Distribution: Known only from Jackson County.

This is a garden ornamental that looks very much like the native *Conoclinium coelestinum*. The only specimen was collected by the author in July 1954, in a vacant lot in Murphysboro, Jackson County. I had originally called it *Eupatorium (Conoclinium) coelestinum*, but when I reexamined the specimen in preparation for this book, I discovered it has cypselae with aristate scales instead of barbellate bristles.

Ageratum conyzoides flowers in July and August in Illinois.

76. **Brickellia** Ell.—False Boneset

Annuals, perennials (in Illinois), or shrubs; stems erect, branched; leaves cauline, simple, mostly alternate, 3-nerved from the base, sometimes glandular-dotted; heads numerous, in corymbs, discoid, not subtended by bractlets; involucre campanulate; phyllaries up to 50, in 3–9 series, glandular-dotted (in Illinois), unequal; receptacle flat or convex, epaleate; ray flowers absent; disc flowers up to 35 (in Illinois), tubular, cream, bisexual, fertile, the corolla 5-lobed; cypselae prismatic, 10-ribbed, glabrous or pubescent; pappus of up to 80 plumose bristles in 1 series (in Illinois) or barbellate bristles.

As considered here, *Brickellia* consists of about 100 species. For many years, our species was placed in the genus *Kuhnia*, but *Brickellia* has been conserved.

Brickellia is very similar to *Eupatorium* but has mostly alternate leaves, cream disc flowers, and up to 80 plumose pappus bristles.

Only the following species occurs in Illinois.

1. **Brickellia eupatorioides** (L.) Shinners, Sida 4:274. 1971.
Kuhnia eupatorioides L. Sp. Pl., ed. 2, 2:1662. 1763.
Kuhnia critonia Willd. Sp. Pl. 3:1773. 1804.
Kuhnia glutinosa Ell. Bot. S. C. & Ga. 2:292. 1823.
Kuhnia suaveolens Fresen. Linnaea 13. Lit. 94. 1839.

Perennial herbs from a woody caudex; stems erect, branched, viscid-pubescent, to 1.5 m tall; leaves mostly alternate, simple, linear-lanceolate to lanceolate to lance-ovate, acute to acuminate at the apex, tapering to the short-petiolate base, to 10 cm long, to 4 cm wide, serrate to nearly entire, more or less pubescent, glandular-dotted, the petioles up to 10 mm long; heads several in a corymb, discoid; involucre campanulate; phyllaries 20–25, in 4–6 series, unequal, striate, scarious, glandular-dotted, the outer lance-ovate to ovate, obtuse to acute to acuminate, the inner lanceolate, obtuse to aristate; receptacle flat to slightly convex, epaleate; ray flowers absent; disc flowers 6–35, tubular, cream-colored, bisexual, fertile, the corolla 5-lobed, 2.7–5.5 mm long; cypselae prismatic, 10-ribbed, glabrous or pubescent; pappus of 20–28 white, usually plumose bristles to 8 mm long.

Brickellia eupatorioides differs from similar-appearing species of *Eupatorium* in its mostly alternate leaves, cream-colored disc flowers, and plumose pappus. It has often been placed in the genus *Kuhnia*.

Three varieties occur in Illinois.

a. Outer phyllaries up to half as long as the inner, obtuse to acute to short-acuminate, not twisted at the tip.
    b. Flowers 6–15 per head; leaves narrowly lanceolate to lance-ovate . . . . 1a. *B. eupatorioides* var. *eupatorioides*
    b. Flowers 15–35 per head; leaves linear-lanceolate to lanceolate . . . . 1b. *B. eupatorioides* var. *corymbulosa*
a. Outer phyllaries half as long as to equaling the inner, long-acuminate and twisted at the apex . . . . 1c. *B. eupatorioides* var. *texana*

1a. **Brickellia eupatorioides** (L.) Shinners var. **eupatorioides**

Leaves narrowly lanceolate to lance-ovate, to 10 cm long, to 4 cm wide; outer phyllaries up to half as long as the inner, acute to acuminate, not twisted at the tip; flowers 6–15 per head, the corolla 4.5–6.0 mm long; cypselae 4.0–5.5 mm long.

Common Name: False boneset.
Habitat: Prairies, dry soil, black oak savannas.
Range: Pennsylvania to Illinois, south to Texas and Florida.
Illinois Distribution: Known from the southwestern corner of the state.

The typical variety has fewer flowers per head and generally wider leaves than the other 2 varieties in Illinois.

This variety flowers from July to October.

1b. **Brickellia eupatorioides** (L.) Shinners var. **corymbulosa** (Torr. & Gray) Shinners, Sida 4:274. 1971.
Kuhnia eupatorioides L. var. *corymbulosa* Torr. & Gray, Fl. N. Am. 2:78. 1841.
Kuhnia eupatorioides L. var. *ozarkana* Shinners, Wrightia 1:136. 1946.

Leaves linear-lanceolate to lanceolate, to 9 cm long, to 1.5 cm wide; outer phyllaries up to half as long as the inner, obtuse to acute to short-acuminate, not twisted at the tip; flowers 15–35 per head, the corolla 4.7–8.0 mm long; cypselae 2.7–5.0 mm long.

Common Name: Narrow-leaved false boneset.
Habitat: Dry prairies; black oak savannas.
Range: Wisconsin to Montana, south to New Mexico, Texas, Arkansas, and Indiana.
Illinois Distribution: Occasional to common in the northern three-fourths of Illinois, much less common elsewhere.

This variety has narrower leaves and more flowers per head than the other 2 varieties in Illinois.

Brickellia eupatorioides var. *corymbulosa* flowers from August to October.

1c. **Brickellia eupatorioides** (L.) Shinners var. **texana** (Shinners) Shinners, Sida 4:274. 1971.
Kuhnia eupatorioides L. var. *texana* Shinners, Wrightia 1:136. 1946.

Leaves linear-lanceolate to lanceolate, to 7 cm long, to 2 cm wide; outer phyllaries half to as long as the inner, long-acuminate and twisted at the tip; flowers 12–24 in a head, the corolla 4.7–7.0 mm long; cypselae 4–5 mm long.

Common Name: Ozark false boneset.
Habitat: Hill prairies (in Illinois).
Range: Illinois to Kansas, south to Texas and Arkansas.
Illinois Distribution: Known only from Jackson County.

This variety, distinguished by the very unequal phyllaries twisted at the apex, was found in the Pine Hills Annex, one-half mile north of the Jackson–Union county line, where it occurred in a hill prairie.

Brickellia eupatorioides var. *texana* flowers from July to October.

77. **Liatris** Gaertn. ex Scribn. *nomen conserv.*–Gayfeather; Blazing-star

Perennial herbs from a cormlike base; stems erect, mostly unbranched; leaves simple, alternate, mostly narrow, usually 1-nerved, entire, glandular-dotted; heads several to numerous in corymbs, racemes, or spikes, discoid, not subtended by bractlets; involucre campanulate or hemispheric; phyllaries up to 40 in 3–7 series, unequal; receptacle flat, epaleate; ray flowers absent; disc flowers up to 85, tubular, mostly purple, bisexual, fertile, the corolla 5-lobed; cypselae prismatic, 8- to 10-ribbed, pubescent, glandular-dotted; pappus of up to 40 barbellate to plumose bristles.

There are 37 species and several hybrids in the genus, all found in the United States, Mexico, and the West Indies.

Some of our species were described originally in *Lacinaria* or *Serratula*, which are older than *Liatris*, but the generic name *Liatris* has been conserved.

I recognize 10 species and 4 hybrids in Illinois. I find the identification of some of the taxa difficult, because several of them are variable, and hybridization occurs among several species. Gaiser (1946) made a comprehensive study of *Liatris*, and I have used her work and my own field experience to come to the conclusions presented here. Study of the pappus and the phyllaries is necessary for proper identification.

Most species have attractive inflorescences, and several of them are grown as ornamentals. Several species have white-flowered forms. With its purple corymbs, spikes, or racemes and numerous narrow alternate leaves, the genus should not be confused with any other genus in Illinois.

1. Pappus plumose or semiplumose.
  2. Larger leaves 3- or 5-nerved.
    3. Phyllaries equal . . . . 1. *L. squarrosa*
    3. Phyllaries unequal.
      4. Phyllaries appressed.
        5. Stems glabrous; corolla tube glabrous within . . . . 2. *L. cylindracea*
        5. Stems pubescent, at least above; corolla tube pilose within . . . 3. *L. X gladewitzii*
      4. Phyllaries spreading, reflexed, or recurved.
        6. Stems hirsute; cypselae 5.5–6.5 mm long . . . . 4. *L. hirsuta*
        6. Stems slightly pilose; cypselae 3–5 mm long . . . . 5. *L. X ridgwayi*
  2. All leaves 1-nerved . . . . 6. *L. punctata*
1. Pappus barbellate.
  7. Larger leaves 3- or 5-nerved.
    8. Flowers usually 5–8 per head.
      9. Phyllaries obtuse; stems glabrous . . . . 7. *L. spicata*
      9. Phyllaries acute to acuminate; stems usually pubescent . . . . 8. *L. pycnostachya*
    8. Flowers 20–40 per head.
      10. Stems glabrous; phyllaries with a broad scarious margin . . . . .9. *L. X steelei*
      10. Stems usually with some pubescence; phyllaries with a narrow scarious margin . . . . :. 10. *L. X sphaeroidea*
  7. All leaves 1-nerved.
    11. Heads on peduncles 8–50 mm long; flowers 30–80 per head; corolla tube glabrous within . . . . 11. *L. scariosa*
    11. Heads sessile or on peduncles up to 8 mm long; flower 12–40 per head; corolla tube pilose within.
      12. Stems hispid, at least below, or sometimes glabrous; phyllaries with erose margins, some of them reflexed . . . . 12. *L. aspera*
      12. Stems pubescent but not hispid; phyllaries with more or less entire margins, spreading or erect.
        13. Phyllaries and leaves with short, spreading hairs, the phyllaries without a scarious margin, mostly erect; flowers 25–40 per head; basal leaves oblanceolate; pappus 8–10 mm long; cypselae about 5 mm long; corolla tube 10–15 mm long . . . . 13. *L. scabra*
        13. Phyllaries and leaves glabrous or nearly so, the phyllaries with a narrow scarious margin, mostly squarrose; flowers 16–24 per head; basal leaves obovate; pappus 5–6 mm long; cypselae about 4 mm long; corolla tube 7 mm long . . . . 14. *L. squarrulosa*

   1. **Liatris squarrosa** (L.) Michx. Fl. Bor. Am. 2:92. 1803.
Serratula squarrosa L. Sp. Pl. 2:818. 1753.
Lacinaria squarrosa (L.) Hill, Hort. Kew. 70. 1768.
Liatris squarrosa (L.) Michx. f. *alba* Evers & Thieret, Rhodora 59:181. 1957.

Perennial herbs from corms; stems erect, mostly unbranched, glabrous or nearly so, to 80 cm tall; leaves numerous, alternate, rather rigid, broadly linear to lanceolate, to 15 cm long, to 12 mm wide, acute at the apex, tapering to the base, glabrous or somewhat pubescent, the broadest ones 3- or 5-nerved, sparsely glandular-dotted; heads up to 25 in loose racemes or spikes, sessile or on peduncles up to 8 mm long, discoid; involucre 10–15 mm high, 6–10 mm wide, campanulate; phyllaries several in 5–7 series, spreading to reflexed, equal, oblong to ovate, glabrous or pubescent, with a scarious margin, acute to acuminate, ciliate; receptacle flat, epaleate; ray flowers absent; disc flowers 23–45, tubular, bisexual, fertile, the corolla 5-lobed, purple, rarely white, the corolla tube glabrous within; cypselae prismatic, 10-ribbed, pubescent, glandular-dotted, 4.0–5.5 mm long; pappus plumose, up to 12 mm long.

Common Name: Scaly blazing-star.
Habitat: Dry open woods, prairies.
Range: Wisconsin to Illinois to Missouri, south to Louisiana, Florida, and Virginia.
Illinois Distribution: Occasional in southern Illinois.

This species is distinguished by its plumose pappus, more or less equal phyllaries, glabrous stems, and 3- or 5-nerved leaves. A white-flowered form, known as f. *alba*, has been found in Illinois.

Liatris squarrosa flowers from July to September.

2. **Liatris cylindracea** Michx. Fl. Bor. Am. 2:93. 1803.
Lacinaria cylindracea (Michx.) Kuntze, Rev. Gen. Pl. 349. 1891.
Liatris cylindracea Michx. var. *bartelii* Steyerm. Rhodora 59:23–24. 1957.

Perennial herbs form globose corms; stems erect, unbranched or branched above, glabrous, to 70 cm tall; leaves numerous, alternate, rigid, linear-lanceolate, acuminate at the apex, tapering to the sessile base, glabrous or with a few hairs on the lower surface, the broadest ones 3- or 5-nerved, not glandular-dotted, to 20 cm long, to 10 mm wide; heads up to 20 in a raceme, sessile or on peduncles up to 10 mm long, discoid; involucre 2–3 cm high, 1.0–1.5 cm wide, campanulate; phyllaries several in 5–7 series, appressed, unequal, the outer broadly oblong to ovate, glabrous, ciliate, with narrow scarious margins, rounded at the apex, the inner mucronate; receptacle flat, epaleate; ray flowers absent; disc flowers 10–35, tubular, bisexual, fertile, the corolla purple, rarely white, 5-lobed, the lobes pubescent, the tube glabrous within; cypselae prismatic, 10-ribbed, pubescent, not glandular-dotted, 5–7 mm long; pappus plumose, up to 10 mm long.

Common Name: Cylindrical blazing-star.
Habitat: Dry prairies, sand flats, gravel hill prairies.
Range: New York and Ontario to Minnesota, south to Oklahoma and Georgia.
Illinois Distribution: Occasional in the northern three-fourths of Illinois, rare
  elsewhere.

The distinguishing characteristics of this species are its large flowering heads with up to 35 flowers per head, its unequal appressed phyllaries, and its glabrous stems. It is known to hybridize with *L. aspera* Michx. to form *Liatris X gladewitzii*, which is treated next. White-flowered plants occur rarely in the state. They are known as f. *bartelii*.

Liatris cylindracea flowers from July to October.

3. **Liatris X gladewitzii** (Farw.) Farw. ex Shinners, Am. Midl. Nat. 29:37. 1943.
Lacinaria gladewitzii Farw. Am. Midl. Nat. 10:43. 1926.

Perennial herbs from a corm; stems erect, unbranched or branched above, glabrous below, pubescent above, to 90 cm tall; leaves numerous, alternate, rigid, linear-lanceolate, acuminate at the apex, tapering to the sessile base, glabrous above, sparsely pubescent below, the broadest ones 3- or 5-nerved, glandular-dotted, to 25 cm long, to 14 mm wide, the uppermost leaves reduced and narrowly linear; heads 6–9, 1–4 cm apart, in an open raceme, on peduncles 5–10 mm long, discoid; involucre cylindric to turbinate to campanulate, to 18 mm long, to 12 mm wide; phyllaries several in 5–7 series, appressed, unequal, the outer orbicular to ovate, glabrous, ciliate, with purplish scarious margins, rounded at the apex, to 3 mm wide, the inner oblong, glabrous, to 5 mm wide, purplish, the margins erose; receptacle flat, epaleate; ray flowers absent; disc flowers 20–30, tubular, bisexual, fertile, the corolla purple, 5-lobed, the lobes densely pubescent, the tube pilose within; cypselae prismatic, 10-ribbed, pubescent, sparsely glandular-dotted, 6–7 mm long; pappus semiplumose, up to 10 mm long.

Common Name: Gladewitz's blazing-star.
Habitat: Dry soil.
Range: Illinois, Michigan, Wisconsin, Ontario, and Manitoba.
Illinois Distribution: Known from Lake County.

This is the presumed hybrid between *L. cylindracea* and *L. aspera*. It resembles *L. cylindracea* in the narrow, rigid leaves that are glabrous on the upper surface, its open racemes of a few short-pedunculate heads, its appressed phyllaries, and its pubescent corolla lobes. It is similar to *L. aspera* in its pubescent stems, at least above; its narrowly scarious phyllaries; and its corolla tube that is pilose within. The pappus is not as plumose as in *L. cylindracea*, but it is not barbellate either. I call it semiplumose.

This hybrid, which flowers from August to October, has been collected in Lake County.

4. **Liatris hirsuta** Rydb. Brittonia 1:98. 1931.
Liatris squarrosa (L.) Michx. var. *hirsuta* (Rydb.) Greene, Rhodora 48:399. 1946.

Perennial herbs from a corm; stems erect, usually unbranched, hirsute, to 70 cm tall; leaves simple, alternate, narrowly linear-lanceolate to linear, the uppermost linear, acuminate at the apex, tapering to the sessile base, hirsute on both surfaces, the broadest ones 3- or 5-nerved, sparsely glandular-dotted, to 18 cm long, to 7 mm wide; heads in loose racemes or spikes, sessile or on peduncles up to 10 mm long, discoid; involucre campanulate, to 13 mm long, to 10 mm wide; phyllaries several in 5–7 series, spreading to reflexed, unequal, ciliate, oblong, abruptly acute to acuminate, sparsely hirsute, without a scarious margin but ciliate; receptacle flat, epaleate; ray flowers absent; disc flowers 15–30, tubular, bisexual, fertile, the corolla 5-lobed, the lobes hispid, the tube glabrous within; cypselae prismatic, 10-ribbed, pubescent, 5.5–6.5 mm long; pappus plumose, up to 8 mm long.

Common Name: Hirsute blazing-star.
Habitat: Dry woods, cherty slopes.
Range: Illinois to Nebraska, south to Texas and Mississippi.
Illinois Distribution: Scattered in southwestern Illinois.

This species has the general aspect of *L. squarrosa* but has hirsute stems and leaves and unequal phyllaries.

    Liatris hirsuta occurs in dry woods, often on chert, in the southwestern counties of the state. It sometimes has been considered a variety of *L. squarrosa*. In the Pine Hills of Union County, it grows beneath species of *Rhododendron* and *Pinus echinata*.

    Liatris hirsuta flowers from August to October.

    5. **Liatris X ridgwayi** Standl. Rhodora 31:37. 1929.
    Perennial herbs from a corm; stems erect, unbranched or branched above, sparsely pilose, to 75 cm tall; leaves simple, alternate, linear to linear-lanceolate, acuminate at the apex, tapering to the sessile base, usually glabrous above, pubescent below, the largest 3- or 5-nerved, glandular-dotted, to 20 cm long, to 10 mm wide, the uppermost reduced; heads crowded, up to 40 or more, in spikes up to 20 cm long, up to 3 cm wide, sessile or on peduncles up to 6 mm long, discoid; involucre cylindric, 8–12 mm high, 5–10 mm wide; phyllaries several in 5–7 series, unequal, the outer lance-oblong, acuminate at the apex, green or with purple tips, pilose, ciliate, squarrose, the middle and inner oblong, acute to acuminate, purple-tinged, glabrous or nearly so; receptacle flat, epaleate; ray flowers absent; disc flowers about 18 per head, tubular, bisexual, fertile, purple, the corolla 5-lobed, the inner surface of the lobes pubescent; cypselae prismatic, 10-ribbed, pubescent, 3–5 mm long; pappus of numerous plumose bristles up to 9 mm long.

Common Name: Ridgway's blazing-star.
Habitat: Dry soil.
Range: Illinois and Nebraska.
Illinois Distribution: Known from the type specimen collected in Richland County by Robert Ridgway.

This is the presumed hybrid between *L. squarrosa* and *L. pycnostachya*. It resembles *L. pycnostachya* in the number and size of flowering heads. It is similar to *L. squarrosa* in its squarrose outer phyllaries and plumose pappus.

    Liatris X rydgwayi flowers during September and October.

6. **Liatris punctata** Hook. Fl. Bor. Am. 1:306. 1833.

Liatris punctata Hook. var. *nebraskana* Gaiser, Rhodora 48:353–54. 1846.

Perennial herbs from globose or elongated corms; stems erect, usually branched, usually glabrous, to 80 cm tall; leaves alternate, simple, linear, rarely reduced upward, acute at the apex, rounded or truncate at the sessile base, to 15 cm long, to 5 mm wide, 1-nerved, usually glabrous or sparsely hirsute, strongly glandular-dotted; heads numerous, densely crowded in spikes, sessile or nearly so, discoid; involucre cylindric or campanulate, 1.5–2.0 cm high, 7–12 mm wide; phyllaries several in 4–5 series, unequal, oblong to narrowly ovate, acute or obtuse at the apex, with or without cilia, glabrous, glandular-dotted, without a scarious margin; receptacle flat, epaleate; ray flowers absent; disc flowers 4–8 per head, tubular, bisexual, fertile, purple, the corolla 5-lobed, the tube glabrous within; cypselae prismatic, 10-ribbed, pubescent, 5.5–8.5 mm long; pappus of numerous plumose bristles up to 10 mm long.

Common Name: Dotted blazing-star; plains blazing-star.
Habitat: Along a railroad (in Illinois).
Range: Michigan to Alberta, south to New Mexico and Louisiana; adventive in
   Illinois.
Illinois Distribution: Collected in Lisle, DuPage County, in 1925 by A. J. Prisc and not
   seen in Illinois since.

The leaves of this species are heavily and conspicuously glandular-dotted. Our specimen has few cilia along the margin of the phyllaries and has been called var. *nebraskana*.

This is the only species of *Liatris* in Illinois that has plumose pappus and 1-nerved leaves.

Liatris punctata flowers from August to October.

7. **Liatris spicata** (L.) Willd. Sp. Pl. 3:1636. 1803.

Serratula spicata L. Sp. Pl. 2:819. 1753.

Liatris spicata (L.) Willd. f. *albiflora* Britt. Bull. Torrey Club 17:124. 1890.

Lacinaria spicata (L.) Kuntze, Rev. Gen. Pl. 349. 1891.

Perennial herbs from slightly elongated corms; stems erect, mostly un-branched, glabrous, to 1.5 m tall; leaves alternate, simple, linear to linear-lanceo-late to oblanceolate, acuminate at the apex, tapering to the sessile base, glabrous, sparsely glandular-dotted, the lowermost to 35 cm long, to 2 cm wide, 3- or 5-nerved, the uppermost reduced; heads several in a dense spike to 50–(60) cm long, sessile or nearly so, discoid; involucre turbinate to cylindric to campanulate, 4–8 mm wide; phyllaries several in 3–5 series, unequal, oblong to ovate, obtuse at the apex, usually glabrous, ciliate, viscid, with a scarious margin, sparsely glandular-dotted; receptacle flat, epaleate; ray flowers absent; disc flowers 5–15 per head, tubular, bisexual, fertile, purple or rarely white, the corolla 5-lobed, the tube glabrous within; cypselae prismatic, 10-ribbed, pubescent, 3.5–6.0 mm long; pappus of barbellate bristles 5–7 mm long.

Common Name: Marsh blazing-star.
Habitat: Prairies, calcareous springs, fens, sand flats, wet meadows, boggy areas.
Range: Quebec and Ontario to Wisconsin, south to Arkansas, Louisiana, and
Georgia.
Illinois Distribution: Scattered throughout Illinois.

Although similar in appearance to *L. pycnostachya*, *Liatris spicata* differs in its
glabrous stems and obtuse phyllaries. White-flowered plants have been seen in
Illinois. These may be known as f. *albiflora*.
Liatris spicata flowers from July to September.

8. **Liatris pycnostachya** Michx. Fl. Bor. Am. 2:91. 1803.
Liatris bebbiana Rydb. Brittonia 1:99. 1932.

Perennial herbs from globose corms; stems erect, mostly unbranched, glabrous
below, usually puberulent near the top, to 1.5 m tall; leaves alternate, simple,
linear to linear-lanceolate, acuminate at the apex, tapering to the sessile base, glabrous, somewhat glandular-dotted, the lowermost to 20 cm long, to 10 mm wide,
3- or 5-nerved, the uppermost reduced; heads several, crowded in spikes to 30 cm
long, to 3.5 cm wide, short-cylindric, sessile, discoid; involucre campanulate to
cylindric, 8–12 mm high, 6–10 mm wide; phyllaries several in 4–5 series, unequal,
broadly linear to obovate, acute at the apex, glabrous or nearly so, ciliate, without a scarious margin, reflexed or spreading; receptacle flat, epaleate; ray flowers
absent; disc flowers 5–8 per head, tubular, bisexual, fertile, purple, the corolla
5-lobed, the tube glabrous within; cypselae prismatic, 10-ribbed, pubescent, 4–6
mm long; pappus of numerous barbellate bristles 3.5–4.0 mm long.

Common Name: Prairie blazing-star.
Habitat: Prairies, marly fens, gravel hill prairies.
Range: New York to North Dakota, south to Texas and Mississippi.
Illinois Distribution: Common throughout the state.

This species, with its attractive purple spikes, is distinguished from the similar-
appearing *L. spicata* by its pubescent stems and acute phyllaries with reflexed
or spreading tips. The type specimen for *Liatris bebbiana*, which is the same as *L.
pycnostachya*, was collected by M. S. Bebb at Fountaindale, Winnebago County.
This is probably the most common species of *Liatris* in Illinois.
Liatris pycnostachya flowers from July to October.

9. **Liatris X steelei** Gaiser, Rhodora 48:227–28. 1946.
Perennial herbs from globose corms; stems erect, branched or sparingly
branched, glabrous, to 85 cm tall; leaves alternate, simple, lanceolate to linear-
lanceolate, acuminate at the apex, tapering to the sessile base, the lowermost
3- or 5-nerved, to 30 cm long, to 2 cm wide, glabrous or nearly so, the uppermost
reduced; heads 20–25, turbinate, in a spike up to 40 cm long, sessile or short-
pedunculate, discoid; involucre turbinate, 10–15 mm high, 10–12 mm wide;

phyllaries several in 4–5 series, unequal, the outer oblong to ovate, obtuse at the apex, glabrous, the middle and inner linear-oblong, obtuse, glabrous, with conspicuous scarious margins; receptacle flat, epaleate; ray flowers absent; disc flowers 20–25 per head, tubular, bisexual, fertile, purple, the corolla 5-lobed, the tube pilose within; cypselae prismatic, 10-ribbed, pubescent, 4.5–5.5 mm long; pappus of numerous barbellate bristles to 6 mm long.

Common Name: Steele's blazing-star.
Habitat: Sandy soil.
Range: Illinois, Indiana, and Kentucky.
Illinois Distribution: Known from Lake and possibly Cook County.

This is presumed to be the hybrid between *L. spicata* and *L. aspera*. It is similar to *L. spicata* but has larger heads, many more flowers per head, and broader phyllaries with scarious margins.

This hybrid flowers from August to October.

10. **Liatris X sphaeroidea** Michx. Fl. Bor. Am. 2:92. 1803.
Perennial herbs from globose corms; stems erect, mostly unbranched, glabrous or scabrous, to 1 m tall; leaves alternate, simple, linear-lanceolate, to 15 cm long, to 10 mm wide, acuminate at the apex, tapering to the sessile base, pubescent, scabrous, 3- or 5-nerved, the uppermost reduced and linear; heads 20–40 in a raceme or panicle up to 75 cm long, subglobose, sessile or short-pedunculate, discoid; phyllaries campanulate to hemispheric; phyllaries several in 4–5 series, unequal, appressed, the outer obovate to oblong with a scarious margin, more or less crispate along the margin, usually glabrous, the middle and inner more elongated; receptacle flat, epaleate; ray flowers absent; disc flowers 25–40 per head, tubular, bisexual, fertile, purple, the corolla 5-lobed, the tube pilose within; cypselae prismatic, 10-ribbed, pubescent, 4–5 mm long; pappus of numerous barbellate bristles 7–8 mm long.

Common Name: Hybrid blazing-star.
Habitat: Prairie dry soil.
Range: Ontario and Manitoba to Minnesota, south to Arkansas and Tennessee; Nebraska.
Illinois Distribution: Scattered in Illinois.

This is presumed to be the hybrid between *L. aspera* and *L. scariosa* var. *nieuwlandii*. It flowers from August to October.

11. **Liatris scariosa** (L.) Willd. var. **nieuwlandii** (Lunell) E. G. Voss, Mich. Bot. 34:139. 1995.
Laciniaria scariosa (L.) Willd. var. *nieuwlandii* Lunell, Am. Midl. Nat. 2:176. 1912.
Laciniaria scariosa (L.) Willd. var. *subcymosa* Lunell, Am. Midl. Nat. 2:177. 1912.
Laciniaria scariosa (L.) Willd. var. *strictissima* Lunell, Am. Midl. Nat. 2:190. 1912.
Liatris novae-angliae var. *nieuwlandii* (Lunell) Shinners, Am. Mids. Nat. 29:31. 1943.
Liatris nieuwlandii (Lunell) Gaiser, Rhodora 48:325. 1946.

Perennial herbs from subglobose corms; stems erect, mostly unbranched, puberulent, to 1 m tall; leaves alternate, simple, narrowly lanceolate to oblanceolate, acute at the apex, tapering to the sessile base, 1-nerved, to 40 cm long, to 50 mm wide, glabrous or puberulent, glandular-dotted, the uppermost nearly as large; heads up to 20, subglobose, in racemes, discoid, on stout peduncles 8–50 mm long; involucre campanulate to turbinate, 15–25 mm across; phyllaries several in 4–5 series, unequal, oblong to ovate, obtuse at the apex, glabrous or puberulent, with a narrow scarious margin, erect or spreading; receptacle flat, epaleate; ray flowers absent; disc flowers 30–80 per head, tubular, bisexual, fertile, purple, the corolla 5-lobed, the tube glabrous within; cypselae prismatic, 10-ribbed, pubescent, 4.5–6.0 mm long; pappus of numerous barbellate bristles up to 9 mm long.

Common Name: Nieuwland's blazing-star; savanna blazing-star.
Habitat: Savannas, prairies.
Range: New York to Wisconsin, south to Arkansas and West Virginia.
Illinois Distribution: Occasional in central Illinois; also Cook, DuPage, and Will counties.

This is the only *Liatris* in Illinois with barbellate pappus bristles, pedunculate flowering heads, and 1-nerved leaves. I do not believe that typical var. *scariosa* occurs in Illinois, although it was used by several Illinois botanists in the past for *L. aspera*. Plants with slightly narrower involucres and with the upper leaves petiolate, the type of which was collected in Cook County in 1876 by W. W. Calkins, was called *Laciniaria scariosa* var. *subcymosa*. Other plants with slightly narrower involucres and with the upper leaves sessile were called *Laciniaria scariosa* var. *strictissima*. They are known from Cook and Peoria counties. Neither of these last two varieties has been transferred to the genus *Liatris*.

This plant is sometimes considered to be a hybrid. It is occasionally and erroneously called *Liatris ligulistylis* (Nels.) K. Schum., which is a different species not known from Illinois.

This taxon flowers from July to October.

12. **Liatris aspera** Michx. Fl. Bor. Am. 2:92. 1803.
Perennial herbs from globose corms; stems erect, usually unbranched, pubescent throughout or only near the base, to 1.75 m tall; leaves alternate, simple, linear-lanceolate to lanceolate, acute at the apex, tapering to the sessile base, to 25 cm long, to 25 mm wide, 1-nerved, short-pubescent or nearly glabrous, scabrous, glandular-dotted; heads numerous in spikes, subglobose, sessile or on peduncles up to 5 (–8) mm long, discoid; involucre campanulate to turbinate, 15–25 mm across; phyllaries several in 4–5 series, unequal, oblong to obovate, obtuse at the apex, glabrous or nearly so, eciliate, with broad, scarious, erose margins, some of them reflexed; receptacle flat, epaleate; ray flowers absent; disc flowers 12–24 per head, tubular, bisexual, fertile, purple, rarely white, the corolla 5-lobed, the tube pilose within; cypselae prismatic, 10-ribbed, pubescent, 4–6 mm long; pappus of numerous barbellate bristles 7–8 mm long.

The distinguishing characteristics of *L. aspera* are its barbellate pappus, 1-nerved leaves, heads nearly sessile, and erose, reflexed phyllaries.

Two varieties occur in Illinois.

a. Stems and leaves short-pubescent throughout . . . . .12a. *L. aspera* var. *aspera*

a. Lower part of the stems and all the leaves glabrous or nearly so . . . .
. . . . 12b. *L. aspera* var. *intermedia*

12a. **Liatris aspera** Michx. var. **aspera**
Liatris scariosa (L.) Willd. f. *benkei* Macbr. Publ. Field Mus. Nat. Hist., Bot. Ser. 4:124. 1927.
Liatris aspera Michx. f. *benkei* (Macbr.) Fern. Rhodora 51:104. 1949.

Stems and leaves short-pubescent throughout.

Common Name: Rough blazing-star.
Habitat: Prairies, gravel hill prairies, black oak savannas.
Range: Ontario to North Dakota, south to Texas and Florida.
Illinois Distribution: Occasional to common throughout Illinois.

White-flowered plants in Illinois have been called f. *benkei*.
The typical variety flowers from July to November.

12b. **Liatris aspera** Michx. var. **intermedia** (Lunell) Gaiser, Rhodora 48:305. 1946.
Lacinaria scariosa (L.) Willd., var. *intermedia* Lunell, Am. Midl. Nat. 2:177. 1912.

Lower part of the stems and all the leaves glabrous or nearly so.

Common Name: Blazing-star.
Habitat: Prairies, calcareous sand prairies, black oak savannas.
Range: Ontario to Wisconsin, south to Missouri, Texas, and Florida.
Illinois Distribution: Occasional throughout Illinois.

This variety flowers from July to November.

13. **Liatris scabra** (Greene) K. Schum. Just's Bot. Jahresber. 29:569. 1903.
Lacinaria scabra Greene, Pittonia 4:317–18. 1901.

Perennial herbs from globose corms; stems erect, mostly unbranched, pubescent, to 1.5 m tall; leaves alternate, simple, the lowest ones oblanceolate, acuminate at the apex, tapering to the sessile base, 1-nerved, to 25 cm long, to 40 mm wide, short-cylindric to turbinate, pubescent, scabrous, scarcely glandular-dotted; heads several in racemes or spikes, sessile or on peduncles to 8 mm long, discoid; involucre turbinate, 15–20 mm high, 15–20 mm across; phyllaries several in 4–5 series, unequal, oblong to obovate, obtuse to acute at the apex, short-pubescent, without a scarious margin, more or less erect; receptacle flat, epaleate; ray flowers absent; disc flowers 25–40 per head, tubular, bisexual, fertile, purple, the corolla 5-lobed, the tube 10–15 mm long, pilose within; cypselae prismatic, 10-ribbed, pubescent, about 5 mm long; pappus of numerous barbellate bristles 8–10 mm long.

Common Name: Scabrous blazing-star.
Habitat: Prairies, open woods.
Range: Indiana to Missouri, south to Texas and Georgia.
Illinois Distribution: Occasional in the southern two-fifths of the state.

This species is similar to *L. squarrulosa* and often combined with that species. I consider it to be a different species by its more pubescent stems and leaves, more cylindrical heads, more flowers per head, erect nonscarious phyllaries, and longer corolla tubes, cypselae, and pappi. The type locality for this species is in the Pine Hills of Union County, where it was collected by F. S. Earle on September 23, 1890.

It is apparently confined to dry woods and prairies in the southern part of the state.

Liatris scabra flowers in September and October, usually later than *L. squarrulosa*.

14. **Liatris squarrulosa** Michx. Fl. Bor. Am. 2:92. 1803.

Perennial herbs from globose corms; stems erect, mostly unbranched, glabrous or puberulent, to 1.2 m tall; leaves alternate, simple, the basal ones obovate, 1-nerved, acute at the apex, tapering to the sessile base, to 25 cm long, to 50 mm wide, glabrous or nearly so, scarcely scabrous, sparsely glandular-dotted; heads several, in racemes or spikes, subglobose, sessile or on peduncles up to 8 mm long, discoid; involucre campanulate, to 15 mm high, to 15 mm across; phyllaries several in 4–5 series, unequal, oblong to obovate, obtuse or acute at the apex, glabrous or nearly so, ciliate, with a narrow scarious margin; receptacle flat, epaleate; ray flowers absent; disc flowers 16–24 per head, tubular, bisexual, fertile, the corolla 5-lobed, the tube to 7 mm long, pilose within; cypselae prismatic, 10-ribbed, pubescent, about 4 mm long; pappus of numerous barbellate bristles 5–6 mm long.

Common Name: Glade blazing-star.
Habitat: Dry woods, glades.
Range: Illinois to Missouri, south to Arkansas and Tennessee.
Illinois Distribution: Known from Alexander and Union counties.

As considered here, this species is known primarily from cherty glades in Alexander and Union counties. It differs from *L. scabra* as summarized under *L. scabra*.

Liatris squarrulosa flowers from June to September, earlier than *L. scabra*.

78. **Fleischmannia** Sch.-Bip.—Pink Thoroughwort

Annual or perennial herbs; stems erect, branched or unbranched, pubescent; leaves simple, opposite, 3-nerved from the base, more or less glandular-dotted; heads several to numerous in corymbs, discoid, without bristles at the base; involucre hemispheric or campanulate; phyllaries up to 30, in 2–4 series, distinctly nerved, unequal; receptacle flat or conical, epaleate; ray flowers absent; disc flowers up to 50 per head, tubular, bisexual, fertile, usually pink, the corolla 5-lobed; cypselae prismatic, 5-ribbed, glabrous or pubescent; pappus of up to 40 barbellate bristles in 1 series.

This genus comprises 80 species, all native to the New World.

Our species at one time was placed in *Eupatorium*, but it differs in its unequal phyllaries and its pink flowering heads.

1. **Fleischmannia incarnata** (Walt.) R. M. King & H. Robins. Phytologia 19:203. 1970.
Eupatorium incarnatum Walt. Fl. Car. 200. 1788.

Perennial herbs from fibrous roots; stems spreading to ascending, to 1.2 m tall, branched, puberulent; leaves simple, opposite, deltoid-ovate, acuminate at the apex, truncate or subcordate at the petiolate base, serrate or crenate, puberulent on both surfaces, to 7 cm long, to 4 cm wide, the petiole up to 3.5 cm long; heads several to numerous in corymbs, some from the axils of the leaves, discoid; involucre campanulate; phyllaries up to 30 in 2–4 series, unequal, acute at the apex, 2-nerved, glabrous or sparsely puberulent, the outer lanceolate, the inner oblong-lanceolate; receptacle more or less flat, epaleate; ray flowers absent; disc flowers 18–20 per head, tubular, bisexual, fertile, pink, the corolla 5-lobed; cypselae prismatic, 5-ribbed, strigose or sometimes nearly glabrous, 2.0–2.8 mm long; pappus of up to 40 barbellate bristles up to 8 mm long.

Common Name: Pink thoroughwort.
Habitat: Wet woods, swamps.
Range: West Virginia to Missouri, south to Texas and Florida.
Illinois Distribution: Rare in the southern one-sixth of the state, absent elsewhere.

This species resembles species of *Eupatorium* except for the pink flowering heads and the unequal phyllaries. In the past is has been known as *Eupatorium incarnatum*. The leaf shape is similar to the leaves of the blue-flowering *Conoclinium coelestinum*.

Fleischmannia incarnata flowers from August to October.

79. **Mikania** Willd.—Climbing Hempweed

Perennial vines; stems twining, branched, terete or 6-angled; leaves simple, opposite, palmately nerved, sometimes glandular-dotted; heads several to numerous, mostly in corymbs, discoid, not subtended by bractlets; involucre cylindric; phyllaries 4 in 2 series, more or less equal; receptacle flat, epaleate; ray flowers absent; disc flowers 4, tubular, bisexual, fertile, pink or white, the corolla 5-lobed; cypselae prismatic, usually 5-ribbed; pappus of up to 60 barbellate bristles in 1–2 series.

There are about 450 species in the genus, all but 9 in the New World.

This is the only viny genus of Asteraceae in Illinois.

1. **Mikania scandens** (L.) Willd. Sp. Pl. 3:1748. 1803.
Eupatorium scandens L. Sp. Pl. 2:836. 1753.

Viny perennials; stems twining or climbing, usually terete, glabrous or nearly so; leaves simple, opposite, triangular, acute at the apex, cordate at the petiolate base, repand or rarely entire, glabrous or nearly so, palmately nerved, to 12 cm long, to 6 cm wide, the slender petioles up to 4 cm long; heads several to many in corymbs, discoid, pedunculate; involucre cylindric; phyllaries 4 in 1–2 series, often pink, equal, linear to lanceolate, acuminate at the apex, glabrous or puberulent, 5–6 mm long; receptacle flat, epaleate; ray flowers absent; disc flowers 4, tubular, bisexual, fertile, pink, the corolla 5-lobed, 3.5–5.5 mm long; cypselae prismatic, 5-ribbed, densely glandular-dotted, 1.8–2.2 mm long; pappus of numerous white or pink barbellate bristles 4.0–4.5 mm long.

Common Name: Climbing hempweed.
Habitat: Low woods, swamps, banks of streams.
Range: New York to Illinois, south to Texas and Florida.
Illinois Distribution: Southern Illinois; also Kankakee County.

This is the only vine in Illinois with opposite, simple leaves that have repand teeth and a cordate base. It often climbs high in trees in the southern tip of the state.
   Mikania scandens flowers from July to October.

80. **Ageratina** Spach.—White Snakeroot

Perennial herbs (in Illinois) or shrubs; stems erect, branched; leaves simple, opposite, 3-nerved from the base, usually serrate or crenate, glabrous or pubescent; heads numerous in corymbs, discoid, not subtended by bractlets; involucre campanulate; phyllaries up to 30 in 2 series, all equal; receptacle convex, epaleate; ray flowers absent; disc flowers up to 60 per head, tubular, bisexual, fertile, usually white, the corolla 5-lobed; cypselae usually prismatic, 5-ribbed, usually pubescent, glandular-dotted; pappus of numerous barbellate bristles in 1 series.
   About 250 species are in this genus, all native to the New World.
   Most species in the genus have been placed at one time or another in the genus *Eupatorium*, but the phyllaries in *Ageratina* are all the same length, whereas those in *Eupatorium* are of different lengths.

1. **Ageratina altissima** (L.) R. M. King & H. Robins. Phytologia 19:212. 1970.
Ageratum altissimum L. Sp. Pl. 2:839. 1753, *non Eupatorium altissimum* L. (1753).
Eupatorium urticaefolium Reichard, Syst. Pl. 3:719. 1780.
Eupatorium ageratoides L. f. Suppl. 355. 1781.
Eupatorium rugosum Houtt. Nat. Hist. 2:550. 1759.

Perennial herbs from rhizomes; stems erect, branched, glabrous or tomentellous, occasionally glaucous, to 1.5 m tall; leaves simple, opposite, ovate to ovate-lanceolae, acuminate at the apex, rounded or truncate at the base, to 15 cm long, to 10 cm wide, coarsely serrate, mostly glabrous, less commonly tomentellous, on petioles to 35 mm long; heads numerous, terminal and from the upper axils, discoid; involucre campanulate; phyllaries up to 30 in 2 series, linear, acute at the apex, equal, 3–5 mm long, glabrous or puberulent; receptacle convex, epaleate; ray flowers absent; disc flowers 15–30 per head, tubular, bisexual, fertile, white, the corolla 5-lobed, the lobes villous; cypselae prismatic, 5-ribbed, glabrous, 1.7–3.0 mm long; pappus of numerous barbellate bristles in 1 series.

Two varieties occur in Illinois.

a. Stems, leaves, and petioles glabrous or nearly so .... 1a. *A. altissima* var. *altissima*
a. Stems, leaves, and petioles tomentellous .... 1b. *A. altissima* var. *tomentella*

1a. **Ageratina altissima** (L.) R. M. King & H. Robins. var. **altissima**
Stems, leaves, and petioles glabrous or nearly so.

Common Name: White snakeroot.
Habitat: Woods, fields.
Range: New Brunswick to Saskatchewan, south to Texas and Florida.
Illinois Distribution: Common throughout the state.

This is the common variety of *A. altissima* found in Illinois, where it occurs in dry woods and fields throughout the state.

For many years, this species was called *Eupatorium rugosum* Houtt., but the equal phyllaries make it a better fit in *Ageratina*. It has been called *Eupatorium urticaefolium* and *Eupatorium ageratoides* by several Illinois botanists in the past.

This is the plant that up until the 1860s was the cause of milk sickness, which usually resulted in the death of the person. One contracted milk silkness by drinking milk from a cow that had eaten this plant. A lady physician from Harrisburg, Illinois, eventually discovered that this species was the cause. Many infants died from milk sickness, as did some adults, including the mother of Abraham Lincoln.

The typical variety flowers from July to October.

1b. **Ageratina altissima** (L.) R. M. King & H. Robins. var. **tomentella** (Robins.)
Mohlenbr. comb. nov. (Basionym: *Eupatorium urticifolium* Reich. var. *tomentellum*
Robins. Proc. Am. Acad. 47:195. 1911.)
Eupatorium rugosum Houtt. var. *tomentellum* (Robins.) Blake, Rhodora 43:557. 1941.

Stems, leaves, and petioles tomentellous.

Common Name: Pubescent white snakeroot.
Habitat: Dry woods.
Range: Michigan and Wisconsin, south to Illinois and Indiana.
Illinois Distribution: Found sparingly throughout the state.

This variety has the same characteristics as the typical variety except that its stems, leaves, and petioles are covered by short-tomentose hairs.

Ageratina altissima var. *tomentella* flowers from July to September.

Tribe Gnaphalieae Cass.

Annuals or perennials (in Illinois); leaves basal and/or cauline and usually alternate, usually entire, often tomentose; heads discoid or disciform, variously arranged; involucre campanulate to hemispheric; phyllaries up to 30 in 1–10 series, equal or unequal, often scarious along the margins and at the apex; receptacle flat or convex or conic, usually epaleate; ray flowers absent; outer flowers sometimes pistillate, more or less raylike, zygomorphic, yellow, pink, or white; inner flowers bisexual, fertile, or only staminate, actinomorphic, the corolla usually 5-lobed, yellow or purplish; cypselae ovoid to obovoid, often flattened, smooth or papillate, usually up to 5-ribbed; pappus of numerous plumose or barbellate bristles, sometimes reduced to scales.

There are approximately 187 genera and 1,240 species in this tribe, found throughout most of the world.

Five genera occur in Illinois.

1. Heads discoid; plants usually unisexual.
   2. Most leaves basal . . . . 81. *Antennaria*
   2. Most leaves cauline . . . . 83. *Anaphalis*
1. Heads disciform; plants usually bisexual.
   3. At least some of the flowers cream or brownish; pappus bristles usually free.
      4. Heads leafy bracted . . . . 84. *Gnaphalium*
      4. Heads not leafy bracted . . . . .82. *Pseudognaphalium*
   3. Flowers yellow or purple; pappus bristles united at the base . . . . 85. *Gamochaeta*

81. **Antennaria** Gaertn.—Pussytoes

Perennial herbs (in Illinois), sometimes with stolons and/or rhizomes; leaves mostly basal, entire, usually strongly nerved, tomentose or lanose on the lower surface; heads borne variously or singly, often in glomerules, discoid, unisexual; involucre campanulate to hemispheric in the staminate flowers, campanulate to cylindric in the pistillate flowers; phyllaries numerous in 3–6 series, unequal, with scarious margins, variously colored; receptacle flat or convex, epaleate; ray flowers absent; staminate disc flowers numerous, tubular, actinomorphic, the corolla 5-lobed; pistillate flowers narrowly tubular; cypselae ellipsoid to ovoid, glabrous or papillate; pappus connate at the base, of up to 20 capillary barbellate bristles.

Antennaria consists of about 45 species, found in several parts of the world.

This genus differs from *Anaphalis* in its leaves being mostly basal.

Five species of *Antennaria* are known from Illinois.

1. Basal leaves prominently 1-nerved or obscurely 3-nerved, usually less than 1.5 cm wide.
   2. Basal leaves obovate, abruptly contracted below the middle to a petiolate base; stolons short, bearing several leaves . . . . .5. *A. howellii*
   2. Basal leaves cuneate-spatulate, rarely obovate, gradually tapering to the sessile base; stolons long, with very few leaves, or rarely the stolons short . . . .4. *A. neglecta*
1. Basal leaves prominently 3- or 5-nerved, usually more than 1.5 cm wide.
   3. Heads several per plant.
      4. Involucre of pistillate flowering heads 7–10 mm high . . . . 3. *A. parlinii*
      4. Involucre of pistillate flowering heads 5–7 mm high . . . . 1. *A. plantaginifolia*
   3. Head solitary . . . . 2. *A. solitaria*

1. **Antennaria plantaginifolia** (L.) Hook. Fl. Bor. Am. 1:330. 1834.
·*Gnaphalium plantaginifolium* L. Sp. Pl. 2:850. 1753.
Antennaria plantaginea R. Br. var. *petiolata* Fern. Proc. Boston Soc. Nat. Hist. 28:242. 1896.
Antennaria plantaginifolia (L.) Hook. var. *petiolata* (Fern.) A. Heller, Muhlenbergia 1:5. 1900.

Perennial dioecious herbs with basal offshoots and long stolons ending in rosettes; stems erect, to 40 cm high, white-tomentose; leaves mostly basal, suborbicular to narrowly obovate, acute and mucronate or rarely subacute at the apex, tapering to the petiolate base, to 7 cm long, to 4 cm wide, strongly 3- or 5- or 7-nerved, entire, gray-pubescent on the upper surface, tomentose and arachnoid on the lower surface, the cauline leaves alternate, linear, to 3 cm long, to 1 cm wide, entire, tomentose; heads either staminate in cymes 1–4 cm across, or pistillate in loose corymbs, discoid; involucre of staminate heads 5–7 mm high, of the pistillate heads 5–7 mm high; phyllaries several, mainly in 3–6 series, unequal, spreading, linear to narrowly oblong, white; receptacle flat to convex, epaleate; ray flowers absent; staminate disc flowers tubular, 2.5–4.0 mm long; pistillate disc flowers tubular, 3–4 mm long, the corolla 5-lobed, often reddish; cypselae ellipsoid to ovoid, 0.8–1.5 mm long, minutely papillate; pappus of capillary, barbellate bristles 2.5–5.5 mm long.

Common Name: Plantain-leaved pussytoes.
Habitat: Woods, old fields, pastures, oak savannas, gravel hill prairies.
Range: New Brunswick to Manitoba, south to Oklahoma, Mississippi, and Florida.
Illinois Distribution: Occasional to common and scattered throughout the state.

This species, along with *A. parlinii* and *A. solitaria*, has leaves with 3, 5, or 7 conspicuous nerves and a width of more than 1.5 centimeters. Plants with basal leaves only up to 2 cm wide and a subacute apex have been called var. *petiolata*. This variety has been found in Illinois.

The following table summarizes the similarities and differences among these three species.

	A. plantaginifolia	*A. solitaria*	*A. parlinii*
stolons	short or long	long	short or long
stems	to 40 cm	to 35 cm	to 35 cm
basal leaves	suborbicular to obovate	obovate to oblong	suborbicular to obovate
	7 cm × 4 cm	7.5 cm × 4.5 cm	7 cm × 4 cm
	3, 5, or 7 nerves	3 or 5 nerves	3 or 5 nerves
	floccose above	floccose above	gray-pubescent to glabrous
cauline leaves	3 cm × 1 cm	1.7 cm × 1.0 cm	4 cm × 1 cm

staminate involucre	5–7 mm high	8–12 mm high	6–9 mm high
heads	several	1	several
pistillate involucre	5–7 mm	8–14 mm	7–12 mm
staminate flower	2.5–4.0 mm long	4.0–5.5 mm long	3.5–5.0 mm long
pistillate flower	3–4 mm long	4.5–7.0 mm long	4–7 mm long
cypselae	0.8–1.5 mm long minutely papillate	1–2 mm long papillate	1–2 mm long minutely papillate
pappus	2.5–5.5 mm long	4–9 mm long	4–8 mm long

Antennaria plantaginifolia flowers during April and May.

2. **Antennaria solitaria** Rydb. Bull. Torrey Club 24:304. 1897.
Antennaria plantaginifolia (L.) Hook. var. *monocephala* Torr. & Gray, Fl. N. Am. 2:431. 1838.

Perennial dioecious herbs with elongated stolons terminated by a rosette of leaves; stems to 35 cm high, white-tomentose; leaves mostly basal, obovate to oblong, obtuse and apiculate at the apex, tapering into the petiolate base, to 7.5 cm long, to 4.5 cm wide, strongly 3- or 5-nerved, entire, gray-pubescent on the upper surface, tomentose on the lower surface, the cauline leaves alternate, linear, entire, to 5 cm long, to 1 cm wide; head unisexual, solitary; staminate involucre 8–12 mm high; pistillate involucre 8–14 mm high; phyllaries several in 3–6 series, unequal, spreading, linear to narrowly oblong, white at the apex; receptacle flat to convex, epaleate; ray flowers absent; staminate disc flowers tubular, 4.0–5.5 mm long; pistillate disc flowers 4.5–7.0 mm long, the corolla 5-lobed, sometimes reddish; cypselae ellipsoid to ovoid, 1–2 mm long, papillate; pappus of capillary, barbellate bristles 4.5–9.0 mm long.

Common Name: One-headed pussytoes.
Habitat: Rich woods.
Range: Pennsylvania to Illinois, south to Louisiana and Georgia.
Illinois Distribution: Known only from Hooven Hollow, Hardin County.

This species is readily recognized by its solitary flowering head. Its leaves are strongly 3- or 5-nerved. The pubescence on the upper surface of the leaves is usually sparser than in *A. plantaginifolia*. The flowering head is also larger than the heads of either *A. plantaginifolia* or *A. parlinii*.

The only location for this species in Illinois is where the first collection of *Micranthes virginiana* was found. When John Voigt and I first found this plant during the 1950s, we thought it was an immature specimen of *A. plantaginifolia*.

Antennaria solitaria flowers during April and May.

3. **Antennaria parlinii** Fern. Gard. & Forest 10:184 1897.

Perennial dioecious herbs with short to long stolons terminated by a rosette of leaves; stems erect, to 35 cm tall, sometimes with purplish glands; leaves mostly basal, obovate to suborbicular, obtuse and usually mucronate at the apex, tapering to the petiolate base, to 7 cm long, to 4 cm wide, entire, gray-pubescent to nearly glabrous on the upper surface, tomentose on the lower surface, the cauline leaves alternate, linear, entire, to 4 cm long, to 1 cm wide; heads several in a crowded corymb, unisexual, discoid; staminate involucre 6–9 mm high; pistillate involucre 7–12 mm high; receptacle flat to convex, epaleate; phyllaries many in 3–6 series, linear to narrowly oblong, white at the apex; staminate heads 3.5–5.0 mm high; pistillate heads 4–7 mm high, the corolla 5-lobed, usually reddish; cypselae ellipsoid to obovoid, 1–2 mm long, minutely papillate; pappus of capillary, barbellate bristles 4–8 mm long.

Two subspecies occur in Illinois.

a. Upper surface of leaves glabrous or nearly so from the beginning; upper part of stem usually with purple glands . . . . 3a. *A. parlinii* ssp. *parlinii*
a. Upper surface of leaves arachnoid from the beginning, tardily glabrate with age; upper part of stem without purple glands . . . . .3b. *A. parlinii* ssp. *fallax*

3a. **Antennaria parlinii** Fern. ssp. **parlinii**

Antennaria arnoglossa Greene, Pittonia 3:321. 1898.
Antennaria parlinii Fern. var. *arnoglossa* (Greene) Fern. Proc. Boston Soc. Nat. Hist. 28:243. 1898.
Antennaria plantaginifolia (L.) Hook. var. *arnoglossa* (Greene) Cronq. Rhodora 47:183. 1945.

Upper surface of leaves glabrous or nearly so from the beginning; upper part of plant usually with purple glands.

Common Name: Parlin's pussytoes.
Habitat: Open woods, fields, pastures, savannas, sand prairies.
Range: Nova Scotia to Manitoba, south to Iowa, Arkansas, Texas, and Georgia.
Illinois Distribution: Occasional throughout the state.

The purple glands on the stems are distinctive, as are the nearly glabrous leaves. This subspecies flowers from April to June.

3b. **Antennaria parlinii** Fern. ssp. **fallax** (Greene) R. J. Bayer & Stebbins, Syst. Bot. 7: 310. 1982.

Antennaria fallax Greene, Pittonia 3:321. 1898.
Antennaria arnoglossa Greene var. *ambigens* Greene, Pittonia 3:320. 1898.
Antennaria occidentalis Greene, Pittonia 3:322. 1898.
Antennaria calophylla Greene, Pittonia 3:347–48. 1898.
Antennaria ambigens (Greene) Fern. Rhodora 1:150. 1899.
Antennaria munda Fern. Rhodora 38:229–30. 1936.
Antennaria fallax Greene var. *calophylla* (Greene) Fern. Rhodora 38:320. 1936.

Antennaria plantaginifolia (L.) Hook. var. *ambigens* (Greene) Cronq. Rhodora 47:183. 1945.

Upper surface of leaves arachnoid at first, tardily glabrate in age; upper part of stem without purple glands.

Common Name: Deceitful pussytoes.
Habitat: Fields, open woods, pastures, savannas.
Range: Nova Scotia to Minnesota and South Dakota, south to Texas and Georgia.
Illinois Distribution: Common in the northern half of Illinois, occasional in the southern half.

This is an extremely variable taxon, as the lengthy nomenclature attests to. It differs from var. *parlinii* in its young arachnoid leaves and the absence of purple glands on upper part of the stems. *Antennaria ambigens*, *A. calophylla*, and *A. munda* are synonyms.

This subspecies flowers during May and June.

4. **Antennaria neglecta** Greene, Pittonia 3:173–74. 1897.
Antennaria campestris Rydb. Bull. Torrey Club 24:304. 1897.
Antennaria erosa Greene, Am. Midl. Nat. 2:78. 1911.
Antennaria neglecta Greene var. *campestris* (Rydb.) Steyerm. Rhodora 62:131. 1960.
Antennaria howellii Greene var. *campestris* (Rydb.) B. Boivin, Phytologia 23:59. 1972.

Perennial dioecious herbs with long or short stolons; stems erect, to 25 m tall, tomentose; leaves basal and cauline, the basal oblanceolate to spatulate to obovate, mucronate at the apex, tapering to the sessile base, to 6 cm long, to 1.5 cm wide, entire, tomentose on the lower surface, gray-pubescent but becoming glabrate on the upper surface, 1-nerved; cauline leaves alternate, linear, to 2.5 cm long; heads unisexual, up to 8 in a corymb or raceme, discoid; staminate involucre 2.5–5.0 mm high; pistillate involucre 4.5–7.0 mm high; receptacle flat to convex, epaleate; phyllaries many in 3–6 series, linear, white at the apex, often purplish at the base; staminate heads 3–5 mm long; pistillate heads 6–9 mm long, the corolla 5-lobed; cypselae ellipsoid to obovoid, 1.0–1.5 mm long, minutely papillate; pappus of capillary, barbellate bristles 3.5–9.0 mm long.

Common Name: Field pussytoes.
Habitat: Fields, prairie remnants, open woods, savannas.
Range: Nova Scotia to British Columbia, south to Colorado, Oklahoma, Arkansas, and Virginia.
Illinois Distribution: Common in the northern half of Illinois, occasional in the southern half.

This species is similar to *A. howellii*, differing in its narrower leaves that are sessile.

Some specimens have rather short stolons, and the leaves tend to be more obovate. These have been recognized in the past as *A. campestris* or *A. neglecta* var. *campestris*.

The type for *Antennaria erosa* Greene was collected by E. L. Greene from Sando-val, Marion County, on June 12, 1898.

Antennaria neglecta is a fairly common plant in fields and prairies. It flowers during April and May.

5. **Antennaria howellii** Greene, Pittonia 3:174. 1897.

Perennial herbs with stolons; stems erect, to 35 cm tall tomentose; leaves basal and cauline, the basal spatulate to oblanceolate to obovate, acute at the apex, tapering to the sessile or petiolate base, to 4.5 cm long, to 2.0 cm wide, entire, pubescent or less commonly glabrous on the upper surface, tomentose on the lower surface, 1-nerved; cauline leaves alternate, linear, entire, to 3.5 cm long; heads 3–15 in corymbs, unisexual, discoid; staminate involucre 6.0–6.5 mm high; pistillate involucre 6–11 mm high; phyllaries several in 3–6 series, linear, white or cream at the apex, sometimes rose at the base; receptacle flat to convex, epaleate; staminate flowers 3–4 mm high; pistillate flowers 3.5–8.0 mm high, the corolla 5-lobed; cypselae ellipsoid to ovoid, 0.8–1.7 mm long, minutely papillate; pappus of capillary, barbellate bristles 4–9 mm long.

Three subspecies may be recognized in Illinois. Typical ssp. *howellii*, which has some of the basal leaves 3-nerved, has not been found in the state.

a. Upper surface of the leaves glabrous or nearly so . . . . 5a. *A. howellii* ssp. *canadensis*
a. Upper surface of the leaves pubescent.
  b. Basal leaves mostly spatulate, petiolate . . . . 5b. *A. howellii* ssp. *neodioica*
  b. Basal leaves mostly obovate, sessile . . . . 5c. *A. howellii* ssp. *petaloidea*

5a. **Antennaria howellii** Greene ssp. **canadensis** (Greene) R. J. Bayer, Brittonia 41:397. 1989.
Antennaria canadensis Greene, Pittonia 3:275 1898.

Upper surface of leaves glabrous or nearly so.

Common Name: Smooth Howell's pussytoes.
Habitat: Woods.
Range: Labrador to Yukon, south to Minnesota, Illinois, and Virginia.
Illinois Distribution: Known only from Lake County.

There is a specimen from Lake County in which the upper surface of the leaves is nearly glabrous. I am considering this to be ssp. *canadensis*.

This subspecies flowers in May.

5b. **Antennaria howellii** Greene ssp. **neodioica** (Greene) R. J. Bayer, Brittonia 41:397. 1989.
Antennaria neodioica Greene var. *attenuata* Fern. Proc. Boston Soc. Nat. Hist. 28:245. 1898.
Antennaria neglecta Greene var. *attenuata* (Fern.) Cronq. Rhodora 47:184. 1945.
Antennaria neglecta Greene var. *neodioica* (Greene) Cronq. Man. Vasc. Pl. N.E. U.S., ed. 2, 863. 1991.

Upper surface of leaves pubescent; basal leaves mostly spatulate, petiolate.

Common Name: Howell's pussytoes.

Habitat: Fields, open woods, black oak savannas.

Range: Newfoundland to British Columbia, south to Oregon, Colorado, Kansas, Tennessee, and North Carolina.

Illinois Distribution: Rare; known from DeKalb, Henry, Kendall, and Knox counties.

The petiolate basal leaves distinguish this subspecies from the others in Illinois.

This subspecies has been called *A. neodioica* in the past. It flowers during May and June.

5c. **Antennaria howellii** Greene ssp. **petaloidea** (Fern.) R. J. Bayer, Brittonia 41:397. 1989.

Antennaria neodioica Greene var. *petaloidea* Fern. Proc. Boston Soc. Nat. Hist. 28:245. 1898.

Antennaria petaloidea (Fern.) Fern. Rhodora 1:73. 1899.

Antennaria neodioica Greene ssp. *petaloidea* (Fern.) R. J. Bayer, Syst. Bot. 7:310. 1982.

Antennaria neglecta Greene var. *petaloidea* (Fern.) Cronq. Man. Vasc. Pl. N.E. U.S., ed. 2, 863. 1991.

Upper surface of leaves pubescent; basal leaves mostly obovate, sessile; stolons short.

Common Name: Petaloid pussytoes.

Habitat: Fields, woods, oak savannas, prairies.

Range: Newfoundland to British Columbia, south to Oregon, Colorado, Illinois, and North Carolina.

Illinois Distribution: Confined to northern Illinois.

This subspecies flowers in May and June.

82. **Pseudognaphalium** Kirpicz.—Cudweed

Annual or biennial herbs (in Illinois), usually with a taproot; stems usually erect, tomentose, sometimes glandular; leaves simple, basal or basal and cauline, entire, usually tomentose on the lower surface; heads disciform, not unisexual, arranged in glomerules, corymbs, or panicles, not subtended by leafy bracts; involucre campanulate to cylindric; phyllaries many, in several series, unequal; receptacle flat, epaleate; ray flowers absent; outer flowers numerous, pistillate, yellowish; inner flowers up to 40, tubular, bisexual, fertile, yellowish; cypselae oblong to cylindric, usually somewhat flattened, glabrous, sometimes with ribs or papillae; pappus of 10–12 capillary, barbellate bristles free at base, caducous.

Approximately 100 species are in the genus. For many years, the species of *Pseudognaphalium* were placed in *Gnaphalium*. The heads in *Pseudognaphalium* are not subtended by leafy bracts as they are in *Gnaphalium*.

Two species occur in Illinois.

1. Leaves not decurrent; stems without stipitate glands; cypselae smooth 1. *P. obtusifolium*
1. Leaves decurrent; stems stipitate-glandular; cypselae papillate . . . . 2. *P. macounii*

1. **Pseudognaphalium obtusifolium** (L.) Hilliard & B. L. Burtt, Bot. Journ. Linn. Soc. 82:205. 1981.
Gnaphalium obtusifolium L. Sp. Pl. 2:851. 1753.
Gnaphalium polycephalum Michx. Fl. Bor. Am. 2:127. 1803.

Biennial aromatic herbs from a taproot; stems erect, branched or unbranched, to 1.2 mm tall, tomentose; leaves simple, alternate, oblong-lanceolate to lanceolate, obtuse to acute at the apex, tapering to the sessile but not decurrent base, entire, to 10 cm long to 10 mm wide, green and nearly glabrous on the upper surface, white and tomentose on the lower surface; heads disciform, not unisexual, in glomerules arranged in corymbs or panicles; involucre campanulate; phyllaries many, in 4–6 series, glabrous or tomentose basally, obtuse at the apex, whitish, the outer oblong or ovate-oblong, the inner narrowly oblong; receptacle flat, epaleate; ray flowers absent; outer flowers numerous, pistillate, fertile; inner flowers up to 10, tubular, bisexual, fertile; cypselae oblong, more or less flattened, glabrous, 1–2 mm long; pappus of 10–12 capillary, barbellate bristles free at base.

Common Name: Sweet everlasting; old field balsam.
Habitat: Fields, pastures, prairies, open woods.
Range: Nova Scotia to Minnesota, south to Texas and Florida.
Illinois Distribution: Common throughout Illinois; in every county.

This species is very fragrant when picked or crushed. It occurs in a wide variety of open habitats. Plants without tomentose stems but with glands have not been found in Illinois, but they are expected since they are known from Indiana, Kentucky, and Missouri. These are known as *P. micradenium* (Weather.) G. L. Nesom.

Many Illinois botanists up until 1936 called this species *Gnaphalium polycephalum*.

Pseudognaphalium obtusifolium flowers from July to October.

2. **Pseudognaphalium macounii** (Greene) Kartesz in Kartesz & Meacham, Synth. N. Am. Fl. 30. 1999.
Gnaphalium macounii Greene, Ottawa Nat. 15:278. 1902.

Biennial aromatic herbs from a taproot; stems erect, up to 75 m tall, branched or unbranched, stipitate-glandular, usually tomentose in the lower half; leaves simple, alternate, linear-lanceolate to linear, acute at the apex, tapering to the decurrent but rarely clasping base, to 10 cm long, to 12 mm wide, entire, often nearly glabrous and green on the upper surface, densely white-tomentose on the lower surface; heads several in glomerules, arranged in corymbs, disciform, not unisexual; involucre campanulate to subglobose; phyllaries many, in 4–5 series, unequal, oblong to ovate, acute at the apex, shiny, yellow-white; receptacle flat, epaleate; ray flowers absent; outer flowers very numerous, pistillate, fertile; inner flowers up to 20, tubular, bisexual, fertile; cypselae oblong, more or less flat, papillate, 1–2 mm long; pappus of 10–12 capillary, barbellate bristles free at the base.

Common Name: Macoun's rabbit tobacco.
Habitat: Sandy soil.
Range: Nova Scotia to British Columbia, south to California, New Mexico,
Tennessee, and Virginia.
Illinois Distribution: Very rare; known only from Clark County.

This species, very rare in Illinois, differs from the very common *P. obtusifolium* in its nearly glabrous but stipitate-glandular stems, its decurrent leaves, and its papillate cypselae.

Pseudognaphalium macounii flowers from July to September.

83. **Anaphalis** DC.—Pearly Everlasting

Perennial herbs (in Illinois) with rhizomes; stems erect, white-tomentose; leaves simple, basal or mostly cauline, the basal, entire, usually white-tomentose on the lower surface, eglandular (in Illinois), the cauline alternate, entire, white-tomentose on the lower surface, green and glabrate or white-tomentose on the upper surface; heads in corymbs or spikes, discoid, unisexual; involucre subglobose; phyllaries many, in 8–12 series, pearly white, unequal, more or less chartaceous; receptacle flat to convex, epaleate; ray flowers absent; outer flowers numerous, pistillate, fertile, yellowish; inner flowers up to 50, staminate, yellowish; cypselae oblongoid to obovoid, 2-nerved, scabrous; pappus of up to 20, free or basally connate capillary, barbellate bristles, caducous.

Anaphalis is a genus of about 110 species, mostly in central Asia and India. One species occurs in North America.

This genus is similar to *Antennaria* in that the flowering heads are usually unisexual. It differs from *Antennaria* in its mostly cauline leaves.

1. **Anaphalis margaritacea** (L.) Benth. & Hook. f. Gen. Pl. 2:303. 1873.
Gnaphalium margaritaceum L. Sp. Pl. 2:850. 1753.
Antennaria margaritacea (L.) Hook. Fl. Bor. Am. 1:329. 1833.
Anaphalis margaritacea (L.) Benth. & Hook. var. *intercedens* Hara, Bot. Mag. Tokyo 40:148. 1938.

Perennial herbs with rhizomes; stems erect, branched, to 1 m tall, densely white-tomentose, eglandular; leaves simple, mostly alternate, cauline, linear to linear-lanceolate, to 12 cm long, to 5 cm wide, acuminate at the apex, tapering to the subclasping and decurrent base, entire, white-tomentose on both surfaces, 1- or 3-nerved; heads discoid, numerous in glomerules in corymbs, unisexual; involucre subglobose; phyllaries many, in 8–12 series, unequal, pearly white, the outer ovate, the inner linear; receptacle convex, epaleate; ray flowers absent; outer flowers numerous, pistillate, fertile, tubular, the corolla 5-lobed; inner flowers up to 50, staminate, yellowish; cypselae oblongoid to obovoid, 0.5–1.0 mm long, 2-nerved, scabrous; pappus of up to 20 free, capillary, barbellate bristles.

Common Name: Pearly everlasting.
Habitat: Woods.

Range: Throughout most of the United States, but native apparently only in the northernmost states and Canada. The Illinois specimen was an introduction.

Illinois Distribution: Known only from Cook County, where it has not been seen in more than 100 years.

Our plant may be called var. *intercedens*. It differs from the typical variety in its white-tomentose upper leaf surface, its leaves that are acuminate at the apex, and its much reduced upper leaves.

The chartaceous pearly white phyllaries account for the common name of pearly everlasting. Several Illinois botanists in the past used this binomial for *Pseudognaphalium obtusifolium* (L.) Hilliard & B. L. Burtt.

Anaphalis margaritacea flowers from July to October.

84. Gnaphalium L.—Cudweed

Annual (in Illinois) herbs with taproots or fibrous roots; stems erect or decumbent, much branched, white-tomentose; leaves simple, mostly cauline, alternate, sessile, entire, usually tomentose; heads disciform, in capitate clusters or spikelike glomerules, bisexual, fertile; involucre campanulate; phyllaries many, in 3–5 series, whitish to brownish, equal or unequal, chartaceous apically; receptacle flat, epaleate; ray flowers absent; outer flowers up to 80, pistillate, bisexual, fertile, purple or whitish; inner flowers up to 7, bisexual, fertile, purple or whitish; cypselae oblongoid, glabrous or papillate; pappus of 8–12 free, capillary, barbellate bristles in 1 series, caducous.

As recognized here, *Gnaphalium* has approximately 40 species native to most parts of the world.

In many earlier works, *Gnaphalium* also included plants now assigned to *Pseudognaphalium* and *Gamochaeta*. As treated here, *Gnaphalium* has pappus bristles free from each other and flowering heads subtended by leafy bracts. *Pseudognaphalium* has pappus bristles free from each other, but the flowering heads are not subtended by leafy bracts. *Gamochaeta* has pappus bristles united at the base.

Only the following species occurs in Illinois.

1. Gnaphalium uliginosum L. Sp. Pl. 2:856. 1753.

Annual herbs with taproots or fibrous roots; stems erect or decumbent, branched from the base, to 75 cm long, more or less tomentose; leaves simple, alternate, cauline, linear to oblanceolate, obtuse or subacute at the apex, tapering to the sessile or short-petiolate base, not decurrent, to 5 cm long, to 3 cm wide, entire, pubescent, usually 1-nerved; heads disciform, solitary or in glomerules, subtended and surpassed by linear bracts, bisexual, fertile; involucre campanulate; phyllaries many, in 3–5 series, oblong, tomentose and brownish at the base, whitish at the apex; receptacle flat, epaleate; outer flowers up to 80, bisexual, fertile, purple or whitish; inner flowers up to 7, bisexual, fertile, purple or whitish; cypselae oblongoid, usually glabrous, 1–2 mm long; pappus of up to 12 free, capillary, barbellate bristles, caducous.

Common Name: Low cudweed.
Habitat: Disturbed soil, pastures, moist woods.
Range: Native to Europe; adventive in the central and northern United States.
Illinois Distribution: Known only from Cook and Lake counties.

The leafy bracts subtending the inflorescence and the free bristles of the pappus are distinctive for this species.

Gnaphalium uliginosum flowers from June to September.

85. **Gamochaeta** Wedd.—Cudweed

Annual or biennial (in Illinois) or perennial herbs with taproots and/or fibrous roots; stems erect or decumbent, often tomentose; leaves simple, basal and cauline, the basal usually withered by flowering time, the cauline alternate, entire, glabrous or tomentose on the upper surface, tomentose on the lower surface; heads disciform, in glomerules in spikes, bisexual; involucre campanulate; phyllaries many, in 3–7 series, unequal, brownish or stramineous or purplish, sometimes chartaceous at the apex; receptacle flat or concave, epaleate; ray flowers absent; outer flowers numerous, pistillate, fertile, yellow or purplish; inner flowers up to 7, bisexual, fertile, yellow or sometimes with a purple tinge; cypselae oblong, more or less flattened, papillate; pappus of several capillary, barbellate bristles united at the base, in 1 series, caducous.

Gamochaeta is distinguished from *Gnaphalium* and *Pseudognaphalium* by the pappus bristles being united at the base and falling as a single unit. The receptacle in *Gamochaeta* is concave during fruiting, and the flowers are in spikes.

Gamochaeta comprises 50 species, all native to the New World. Only the following species occurs in Illinois.

1. **Gamochaeta purpurea** (L.) Cabrera, Bol Soc. Argent. Bot. 9:377. 1961.
Gnaphalium purpureum L. Sp. Pl. 2:854. 1753.

Annual or biennial herbs with taproots or fibrous roots; stems ascending to erect, branched or unbranched, to 70 m tall, tomentose; leaves simple, basal and cauline, the basal spatulate to oblanceolate, obtuse to subacute at the apex, tapering to the sessile base, entire, green and usually glabrous on the upper surface, tomentose on the lower surface, to 6 cm long, to 1.5 cm wide, usually withered at anthesis, the cauline similar but smaller and persistent; heads disciform, in spikes, bisexual, the lowermost heads axillary; involucre campanulate; phyllaries many, in 4–5 series, unequal, purplish or whitish, the outer ovate and acute to acuminate at the apex, the inner lanceolate and acute at the apex; receptacle concave, epaleate; ray flowers absent; flowers 3–4 in a head, bisexual, fertile, purple, at least toward the apex; cypselae oblong, more or less flattened, papillate, 0.5–0.7 mm long; pappus of several capillary, barbellate bristles united at the base and falling as a single unit, caducous.

Common Name: Purple cudweed; early cudweed.
Habitat: Fields, open woods, black oak savannas, sand barrens.

Range: Maine to Ontario to Iowa, south to Texas and Florida; Arizona.

Illinois Distribution: Occasional throughout the state, except for the northeastern counties. It is also known from Kankakee County.

This species is distinctive in its spikelike inflorescence and the pappus that falls as a single unit.

Gamochaeta purpurea flowers from May to July.

Tribe Inuleae—Cass.

Annual or perennial herbs, often with taproots; leaves simple, basal and withering at anthesis, the cauline leaves alternate, entire to serrate to divided; heads usually radiate (in Illinois), borne singly or in various inflorescences, bisexual; involucre hemispheric or campanulate; phyllaries several, in 3–7 series, unequal; receptacle flat or convex, epaleate; ray flowers present (in Illinois), yellow, pistillate, fertile; disc flowers bisexual, fertile, usually yellow, the corolla 4- or 5-lobed; cypselae ellipsoid or prismatic, ribbed or angled; pappus of several basally connate barbellate bristles.

This tribe in the past included plants now considered to be in the tribes Gnaphalieae and Plucheae. Without those tribes, the Inuleae consists of about 40 genera and approximately 500 species. Only the following genus occurs in Illinois.

86. Inula L.—Elecampane

Perennial (in Illinois) herbs with taproots; basal leaves withering at anthesis; cauline leaves simple, alternate, serrate or dentate, often woolly on the lower surface; heads radiate (in Illinois), borne singly or in corymbs, bisexual, sometimes showy; involucre hemispheric or campanulate; phyllaries several, in 3–7 series, unequal, pubescent; receptacle flat or convex, epaleate; ray flowers up to 150, pistillate, fertile, yellow; disc flowers numerous, tubular, bisexual, fertile, the corolla 5-lobed, yellow; cypselae prismatic, 4- or 5-ribbed or angled; pappus of up to 60 basally connate bristles, persistent.

There are about 100 species of *Inula*, all of them native to the Old World. Only the following introduced species has been found in Illinois.

1. **Inula helenium** L. Sp. Pl. 2881. 1753.

Stout perennial herbs from a taproot; stems usually unbranched, to 1.5 m tall, densely pubescent, at least above; leaves simple, basal and cauline, the basal withered by anthesis, ovate, acuminate at the apex, tapering to the petiolate base, to 40 cm long, to 20 cm wide, serrate or denticulate, pubescent on the upper surface, velutinous on the lower surface, the cauline elliptic to lanceolate, acute at the apex, tapering or rounded or cordate at the often clasping base; heads radiate, borne singly or in corymbs, bisexual, showy, up to 4 cm across; involucre hemispheric; phyllaries several, in 3–7 series, the outer lanceolate to ovate, velutinous, the inner linear-lanceolate, pubescent; receptacle flat or convex, epaleate; ray flowers up to 100, linear, pistillate, fertile, yellow; disc flowers numerous, tubular, bisexual, fertile, yellow, about 10 mm long; cypselae ellipsoid, 4- or 5-ribbed or angled, glabrous, 3–4 mm long; pappus up to 60 basally connate bristles in 1 series, persistent.

Common Name: Elecampane.
Habitat: Fields, pastures, roadsides, open woods.
Range: Native to Europe and Asia; adventive and scattered in the United States.
Illinois Distribution: Scattered in Illinois, but not common.

This garden ornamental has the appearance of a sunflower but is readily distinguished by its uppermost clasping leaves.

Inula helenium flowers during July and August.

Tribe Plucheae (Cass. ex Dum.) Anderberg

Herbs (in Illinois), shrubs, or trees; leaves simple, usually cauline, alternate, entire, toothed, or pinnately divided; heads disciform, variously arranged; involucre ovoid, campanulate, or hemispheric; phyllaries several, in 3–6 series, free from each other; receptacle flat to convex, epaleate; ray flowers absent; outer flowers pistillate, fertile, usually pink or purplish; inner flowers tubular, bisexual or staminate, pink or purplish, the corolla 4- or 5-lobed; cypselae various, usually ribbed, sometimes flattened; pappus of numerous bristles or scales in 1 or 2 series.

This tribe consists of 27 genera and 220 species, most of them in tropical regions of the world.

Only the following genus occurs in Illinois.

87. **Pluchea** Cass.—Marsh Fleabane

Herbaceous (in Illinois) or woody plants, often strongly aromatic, with taproots or fibrous roots; leaves simple, cauline, alternate, entire or dentate, usually pubescent; heads disciform, borne in corymbs or panicles; involucre campanulate or hemispheric; phyllaries several, in 3–6 series, unequal; receptacle flat, epaleate; ray flowers absent; outer flower pistillate, fertile, pink, purple, white, or yellow; inner flowers staminate, tubular, pink, purple, white, or yellow, the corolla 4- or 5-lobed; cypselae oblong to cylindric, ribbed, usually pubescent; pappus of free or basally connate bristles in 1 series.

As many as 60 species are in *Pluchea*, mostly in subtropical or tropical regions of the world.

Two species occur in Illinois.
1. Leaves petiolate, membranaceous; phyllaries glabrous . . . . 1. *P. camphorata*
1. Leaves sessile, somewhat fleshy; phyllaries pubescent . . . . 2. *P. odorata*

1. **Pluchea camphorata** (L.) DC. Prodr. 5:452. 1836.

Erigeron camphoratum L. Sp. Pl. 2:864. 1753.
Gynema viscida Raf. Ann. Nat. 15. 1820.
Pluchea petiolata Cass. Dict. Sci. Nat. 42:2. 1826.
Pluchea viscida (Raf.) House, Am. Midl. Nat. 7:129. 1921.

Strongly aromatic annual or perennial herbs with fibrous roots; stems erect, branched, to 1.5 m tall, viscid glandular-pubescent, arachnoid; leaves simple, alternate, cauline, membranaceous, elliptic to lanceolate to narrowly ovate, acute to acuminate at the apex, narrow or rounded at the short-petiolate base, to 15 cm long, to 6 cm wide, dentate or serrate, rarely entire, glandular-pubescent; heads several in cymes or corymbs, disciform; involucre campanulate; phyllaries several, in 3–6 series, cream or purplish, unequal, the outer ovate, opaque, glabrous or sparsely pubescent, the inner lanceolate, longer than the outer, translucent, glandular; receptacle flat, epaleate; ray flowers absent; outer flowers several, pistillate, fertile, rose-purple; inner flower up to 40, staminate, rose-purple; cypselae more or less cylindric, 2.5–4.0 mm long, glabrous or puberulent; pappus of several free bristles, persistent.

Common Name: Camphorweed; stinkweed.
Habitat: Swamps and sloughs; marshes.
Range: Pennsylvania to Kansas, south to Texas and Florida.
Illinois Distribution: Occasional in southern Illinois, extending northward to DeWitt County.

This strongly aromatic species is an inhabitant of open marshes and swampy woods. It differs from *P. odorata* in its petiolate leaves and nearly glabrous phyllaries.
Pluchea camphorata flowers from July to October.

2. **Pluchea odorata** (L.) Ca. in F. Cuv. var. **succulenta** (Fern.) Cronq. Fl. S.E. U.S. 1:175. 1980.
Pluchea purpurascens (Swartz) DC. var. *succulenta* Fern. Rhodora 44:227. 1942.

Strongly aromatic herbs with fibrous roots; stems erect, usually branched, to 75 cm tall, viscid-glandular, not arachnoid; leaves simple, cauline, alternate, somewhat fleshy, elliptic to ovate, acute at the apex, tapering or rounded at the sessile base, to 15 cm long, to 5 cm wide, dentate, glabrous or pubescent; heads several in corymbs, disciform; involucre cylindric to campanulate; phyllaries several, in 3–6 series, usually purplish, unequal, pubescent, the outer ovate, the inner lanceolate; receptacle flat, epaleate; ray flowers absent; outer flowers pistillate, fertile, rose or pinkish; inner flowers 20–34, staminate, rose or pink; cypselae more or less cylindric, 3–4 mm long, glabrous or pubescent; pappus of several free bristles, persistent.

Common Name: Salt marsh fleabane.
Habitat: Ditch (in Illinois).
Range: Maine to North Carolina; apparently adventive in Illinois, Indiana, Michigan, and Ontario.
Illinois Distribution: Known only from Cook County.

This species, native to brackish and salt marshes of the Atlantic Coastal Plain, has been found a few times in the upper Midwest. The Illinois plants were found in the Wolf Lake Conservation Area in Cook County.

Typical var. *odorata*, which is not known from Illinois, has narrower involucres and fewer staminate flowers per head.

This variety flowers during August and September.

Tribe Cynareae—Lam. & DC.

Annual or perennial herbs; leaves simple, basal or cauline and alternate, often spiny; heads discoid, bisexual, fertile, or some of them disciform and pistillate or neutral, borne in corymbs, panicles, or racemes; involucre various; phyllaries usually in 2–5 series, unequal, sometimes spiny, sometimes with basal appendages; receptacle flat or convex, usually epaleate; ray flowers absent, although sometimes the peripheral flowers appear raylike; peripheral flowers, when present, pistillate, often yellow; disc flowers tubular, bisexual, fertile, yellow or white, the corolla 5-lobed; cypselae prismatic, angled, or terete; pappus of simple or barbellate or plumose bristles, scales, or both bristles and scales.

There are about 83 genera and 2,500 species in this tribe, mostly native to the Old World. Ten genera occur in Illinois, but only 10 species in the genus *Cirsium* are native.

Key to the Genera of Cynareae in Illinois

1. Leaves with spine-tipped teeth.
  2. Flowers yellow . . . . 97. *Centaurea*
  2. Flowers purple, pink, pale blue, or white.
    3. Pappus of barbellate bristles; phyllaries in 2 series; stems winged . . .89. *Onopordum*
    3. Pappus of simple bristles, plumose bristles, or a crown of scales; phyllaries in several series; stems usually unwinged.
      4. Pappus a crown of scales; flowers pale blue . . . . 88. *Echinops*
      4. Pappus of bristles; flowers purple, pink, or white.
        5. Pappus of plumose bristles. . . . . .91. *Cirsium*
        5. Pappus of simple bristles . . . . 90. *Carduus*
1. Leaves without spine-tipped teeth.
  6. Outer row of flowers appearing ligulate.
    7. Outer phyllaries entire, never spine-tipped.
      8. Phyllaries coriaceous, yellowish, without a hyaline margin . . . . . 94. *Amberboa*
      8. Phyllaries thin, green, with a hyaline margin . . . . .93. *Acroptilon*
    7. Outer phyllaries fimbriate or laciniate, some of them sometimes spine-tipped.
      9. Involucre 3–4 cm high; pappus 6–12 mm long . . . . 95. *Plectocephalus*
      9. Involucre 1.0–1.5 cm high; pappus 3 mm long or less. . . . . 96. *Carthamus*
  6. Outer row of flowers tubular, not appearing ligulate.
    10. Flowers greenish . . . . 92. *Arctium*
    10. Flowers purple or orange.
      11. Flowers purple . . . . 92. *Arctium*
      11. Flowers orange. . . . . 96. *Carthamus*

88. **Echinops** L.—Globe Thistle

Perennial herbs; stems erect, simple or branched, pubescent, spiny; leaves simple, basal and cauline, alternate, toothed or pinnately divided, spine-tipped, pubescent; inflorescence with many sessile flowers in pedunculate, globose heads, subtended by reflexed, deeply cleft bracts; involucre cylindric, subtended by bristles; phyllaries several, in several series, unequal; receptacle turbinate, with subulate scales; ray flowers absent; disc flower one per head, usually bluish, bisexual, fertile, the corolla 5-lobed; cypselae cylindric, 4-angled, truncate at the apex, long-hairy; pappus of many short scales.

This genus comprises about 120 species, all native to Europe, Asia, and Africa. Some species are grown as ornamentals.

Echinops is distinguished by its globose flowering heads and pappus of short scales.

1. **Echinops sphaerocephalus** L. Sp. Pl. 2:814. 1753.

Perennial herbs to 1.5 m tall; stems erect, usually branched, with arachnoid hairs; leaves simple, basal and cauline, alternate, elliptic to narrowly obovate, to 15 cm long, to 8 cm wide, acute at the apex, shallowly pinnately lobed, the lobes lanceolate, revolute, usually with arachnoid hairs, dentate, spine-tipped, the lowest on short, winged petioles, the upper sessile and clasping; flowers many in a globose head, the head up to 5 cm in diameter, discoid, on stout peduncles, subtended by small, reflexed bracts; involucre cylindric; phyllaries several, in several series, unequal, the outer glandular-dotted, the inner fringed; receptacle turbinate, with subulate scales; ray flowers absent; disc flower 1, tubular, bisexual, fertile, pale blue, the tube up to 14 mm long, the corolla 5-lobed, the lobes 6–7 mm long; cypselae cylindrical, 4-angled, pubescent, up to 10 mm long; pappus a crown of connate scales up to 1.5 mm long.

Common Name: Globe thistle.
Habitat: Disturbed soil.
Range: Native to Europe and Asia; adventive in several states.
Illinois Distribution: Known from the northeastern counties and Coles County.

This ornamental is easily recognized by its globose flowering heads and spinescent thistlelike leaves. It rarely escapes from cultivation.

Echinops sphaerocephalus flowers from July to September.

89. **Onopordum** L.—Scotch Thistle

Biennial herbs; stems erect, branched, with spiny wings; leaves simple, basal and cauline, sessile or with a winged petiole, the lobes spine-tipped; heads solitary or few in corymbs, discoid; involucre usually hemispheric or spherical; phyllaries several, in 2–10 series, more or less unequal, spine-tipped; receptacle flat or convex, epaleate, usually pitted; ray flowers absent; disc flowers tubular, bisexual, fertile, white or purple, the tube slender, the corolla 5-lobed; cypselae cylindric, angular, transversely rugose, glabrous; pappus a ring of numerous barbellate or plumose bristles connate at the base.

There may be as many as 60 species in the genus, all native to Europe and Asia.

This is another genus of prickly thistlelike plants, differing from *Echinops* in its nonglobose flowering heads, from *Carduus* in its lack of setae on the receptacle, from *Cirsium* in its strongly winged stems, from *Carthamus* in its lack of scales on the receptacle, and from some species of *Centaurea* in its lack of disciform peripheral flowers.

1. **Onopordum acanthium** L. Sp. Pl. 2:827. 1753.

Stout biennial herbs; stems erect, branched, silvery-tomentose, to 3 m tall, with wings up to 15 mm wide; leaves simple, basal and cauline, alternate, to 6 cm long, to 4 cm wide, acuminate at the apex, decurrent at the base, coarsely toothed to shallowly pinnatifid, densely tomentose, the teeth or lobes spine-tipped; heads usually 2 or 3 in a short corymb, discoid, on a prickly winged peduncle; involucre spherical; phyllaries several, in 2 series, linear, usually tomentose, spine-tipped; receptacle more or less flat, pitted; ray flowers absent; disc flowers tubular, purple or white, the tube up to 20 mm long, the corolla 5-lobed; cypselae cylindrical, angular, transversely rugose, glabrous, 4–5 mm long; pappus a ring of pink or reddish barbellate bristles up to 9 mm long.

Common Name: Scotch thistle; cotton thistle.
Habitat: Disturbed soil.
Range: Native to Europe and Asia; adventive in several states.
Illinois Distribution: Known only from Champaign and Cook counties. It was first collected in Illinois in 1947. It rarely escapes from cultivation.

This thistlelike species is distinguished by its large purple or white flowering heads, its broadly winged and spiny stems, and its barbellate pink or reddish pappus.

Onopordum acanthium flowers from June to August.

90. **Carduus** L.—Musk Thistle; Plumeless Thistle

Annual, biennial, or perennial herbs; stems erect, usually branched, spiny, winged; leaves simple, basal and cauline, dentate or palmately lobed, spine-tipped; heads solitary or several in a corymb, discoid, on spiny peduncles; involucre cylindric or spherical; phyllaries many, in several series, more or less equal with basal appendages, mostly linear, spine-tipped; receptacle flat or convex, epaleate but bearing scales; ray flowers absent; disc flowers numerous, tubular, white, pink, or purple, bisexual, fertile, the tube elongated, the corolla 5-lobed; cypselae ovoid or obovoid, usually somewhat angled, glabrous; pappus a ring of minutely barbed bristles or scales, usually connate at the base.

There are about 90 species of *Carduus*, native to Europe, Asia, and Africa. Three species are adventive in Illinois.

1. Head solitary, nodding; basal appendages of phyllaries 2–7 mm wide; head 2–7 cm across; pappus bristles 13–28 mm long . . . . .3. *C. nutans*
1. Heads 1 to several, erect; basal appendages of phyllaries up to 1.5 mm wide; heads 1.5–2.5 cm across; pappus bristles 11–18 mm long.
   2. Heads 1.8–2.5 cm across; leaf surface glabrous or with septate hairs . . .1. *C. acanthoides*
   2. Heads 1.5–1.8 cm across; leaf surface tomentose, with nonseptate or septate hairs . . . . 2. *C. crispus*

1. **Carduus acanthoides** L. Sp. Pl. 2:821. 1753.

Robust annual or perennial herbs; stems erect, branched, to 2 m tall, glabrous or with septate hairs, spiny winged, the spines to 8 mm long; leaves simple, basal and cauline, the basal to 30 cm long, with sharp spine-tipped teeth or deeply pinnately divided, glabrous below or with septate hairs on the veins, sparsely pubescent above, tapering to a spiny winged petiole; cauline leaves alternate, progressively smaller, with spine-tipped teeth or lobes, sessile; heads solitary or up to 5 in a corymb, 1.8–2.5 cm across, discoid, on spiny winged peduncles up to 5 (–10) cm long; involucre hemispheric; phyllaries numerous, in several series, mostly spreading, linear, glabrous or ciliolate, with basal appendages up to 1.5 mm wide, the outer spine-tipped; receptacle flat or spine-tipped, epaleate, but bearing scales; ray flowers absent; disc flowers numerous, tubular, purple or sometimes white, bisexual, fertile, the tube up to 20 mm long, the corolla 5-lobed; cypselae obovoid, brown, glabrous, 2.5–3.0 mm long; pappus a ring of bristles 11–13 mm long.

Common Name: Plumeless thistle.
Habitat: Pastures, disturbed soil.
Range: Native to Europe and Asia; adventive in most of the upper parts of the United States.
Illinois Distribution: Occasional in the northern one-sixth of Illinois.

This species differs from *Carduus nutans* in its erect flowering heads and from *C. crispus* in its smaller flowering heads and its sparsely pubescent leaf surfaces.

This is one of the prickliest plants in Illinois. When I was on a collecting trip several years ago with Floyd Swink of the Morton Arboretum, he put on a pair of thick leather gloves to collect a part of this plant for a specimen.

Carduus acanthoides flowers from May to October.

2. **Carduus crispus** L. Sp. Pl. 2:821. 1753.

Robust biennial herbs; stems erect, branched, to 1.5 m tall, villous with septate hairs, winged, very spiny, the spines up to 3 mm long; leaves simple, basal and cauline, the basal to 20 cm long, coarsely toothed or shallowly pinnately lobed, the teeth and lobes spine-tipped, glabrous except for long, curled, septate hairs, tapering to a spiny-winged petiole; cauline leaves alternate, similar but progressively smaller and sessile; heads solitary or up to 5 in a corymb, 1.5–1.8 cm across, discoid, on spiny winged peduncles up to 4 cm long; involucre spherical; phyllaries numerous, in several series, appressed to spreading, narrowly lanceolate, glabrous, ciliolate, with basal appendages 0.5–1.0 mm wide, the outer spine-tipped; receptacle flat or convex, epaleate but bearing scales; ray flowers absent; disc flowers numerous, tubular, purple or occasionally white, bisexual, fertile, the tube up to 16 mm long, the corolla 5-lobed; cypselae obovoid, brown, glabrous, 2.5–3.5 mm long; pappus a ring of bristles 11–13 mm long.

Common Name: Curled thistle; welted thistle.
Habitat: Disturbed soil.
Range: Native to Europe and Asia; adventive in a few, primarily eastern states.
Illinois Distribution: Known only from Jackson County, where it apparently has been extirpated.

This species is a slightly smaller version of *C. acanthoides*. However, its involucre is spherical rather than hemispherical.

The Illinois location is now a residential development.

Carduus crispus flowers from May to October.

3. **Carduus nutans** L. Sp. Pl. 2:821. 1753.
Carduus leiophyllus Petrovic. Add. Fl. Agri. Nyss. 105. 1885.
Carduus nutans L. var. *leiophyllus* (Petrovic) G. A. Mulligan & Frankton, Can. Field-
    Nat. 68:35. 1954.

Robust perennial herbs; stems erect, branched, up to 4 m tall, tomentose or glabrous, winged, spiny, the spines up to 10 mm long; leaves simple, basal and cauline, the basal to 40 cm long, deeply pinnately lobed, the lobes with a spiny tip, glabrous or tomentose, tapering to a winged, spiny petiole; cauline leaves alternate, similar but progressively smaller, sessile; head usually solitary, nodding, 2–7 cm across, discoid, on tomentose and occasionally winged peduncles up to 30 cm long; involucre hemispheric; phyllaries numerous, in several series, some of them spreading to reflexed, lanceolate to ovate, glabrous or tomentose, spine-tipped, with broad appendages up to 7 mm wide; receptacle flat or convex, epaleate but bearing scales; ray flowers absent; disc flowers numerous, tubular, bisexual, fertile, rose-purple, the tube up to 28 mm long, the corolla 5-lobed; cypselae ovoid, brown, glabrous, 4–5 mm long; pappus a ring of bristles up to 25 mm long.

Common Name: Nodding thistle; musk thistle.
Habitat: Pastures, along railroads, roadsides.
Range: Native to Europe; introduced into the United States, where it is expanding its
    range rapidly.
Illinois Distribution: Common throughout the state.

This is the only thistle or thistlelike plant in Illinois with a large, rose-purple, nodding, solitary head. Fifty years ago, it was found mostly along railroads in large cities, but more recently it has become widespread in disturbed soil.

Plants with glabrous stems have been called var. *leiophyllus*.

Carduus nutans flowers from May to November.

91. **Cirsium** Mill.—Thistle

Annual, biennial, or perennial herbs; stems erect, branched or unbranched, sometimes with a narrow spiny wing; leaves simple, basal and cauline, alternate, toothed to pinnately lobed, bristle- or spine-tipped; heads few to numerous, in corymbs, racemes, or panicles, discoid; involucre cylindric to spherical to ovoid; phyllaries many, in up to 20 series, more or less equal, usually spine-tipped, sometimes resinous; receptacle flat to convex, epaleate, but with bristles or setae; ray flowers absent; disc flowers numerous, tubular, bisexual, fertile, pink or purple or rarely white, the tube slender, the corolla 5-lobed; cypselae ovoid, sometimes flattened, glabrous; pappus usually a ring of plumose bristles or plumose scales in 3–5 series.

Cirsium consists of about 200 species, native to North America, South America, Europe, Asia, and North Africa.

Cirsium differs from other thistlelike plants in its pappus that consists of plumose bristles or plumose scales.

Ten species are known from Illinois.

1. Involucre up to 1.5 (–2.0) cm high; plants with creeping rootstocks; many of the heads unisexual . . . . 2. *C. arvense*
1. Involucre more than 1.5 cm high; plants without creeping rootstocks; heads bisexual.
   2. Phyllaries resinous and sticky, not spine-tipped or with spines up to 0.5 mm long . . . . 5. *C. muticum*
   2. Phyllaries neither resinous nor sticky, at least the outer ones spine-tipped.
      3. Leaves, or most of them, decurrent; stems usually winged.
         4. Flowers cream-colored; leaf surface not covered by spines; stems unwinged or narrowly winged . . . . 10. *C. pitcheri*
         4. Flowers purple; leaf surface covered by spines; stems broadly winged. . . . . 1. *C. vulgare*
      3. Leaves not decurrent; stems unwinged.
         5. Stems densely white-tomentose, even at maturity; leaves sparsely tomentose on the upper surface, at least when young.
            6. Cypselae 3–5 mm long; tube of corolla up to 15 mm long; plants with shallow horizontal runners; none of the cauline leaves clasping. . . . . . . .8. *C. flodmanii*
            6. Cypselae 6–7 mm long; tube of corolla up to 28 mm long; plants with deep horizontal runners; some of the cauline leaves clasping . . . . 9. *C. undulatum*
         5. Stems glabrous or pubescent, but not white-tomentose; leaves never tomentose on the upper surface.
            7. Leaves densely white-tomentose on the lower surface.
               8. Involucre 1.5–2.5 cm high; cypselae 3–4 mm long . . . . . 7. *C. carolinianum*
               8. Involucre 2.5–3.5 cm high; cypselae 4–6 mm long.
                  9. Leaves deeply pinnatifid, revolute. . . . . .4. *C. discolor*
                  9. Leaves toothed or shallowly pinnatifid, flat. . . . . 3. *C. altissimum*
            7. Leaves gray-hairy beneath, becoming glabrous, never densely tomentose . . . . 6. *C. hillii*

1. **Cirsium vulgare** (Sav.) Tenore, Fl. Nepol. 5:209. 1835.
Carduus vulgaris Sav. Fl. Pis. 2:241. 1798.

Stout biennial herbs from taproots; stems erect, to 2.5 m tall, branched, winged, spiny, with villous, septate hairs; leaves basal and cauline, the basal present or absent at flowering time, oblanceolate to obovate, to 40 cm long, to 15 cm wide, coarsely toothed to pinnatifid, the teeth and lobes spine-tipped, gray-tomentose on the lower surface, short-spiny on the upper surface, tapering to a winged petiole, the cauline alternate, progressively smaller, the lower decurrent to a winged petiole, the upper sessile; heads 1 to few, in corymbs or panicles, discoid, on spiny, winged peduncles to 5 cm long; involucre hemispheric to campanulate, 3–4 cm high; phyllaries several, in 10–12 series, entire, spine-tipped, the outer linear-lanceolate, the inner linear; receptacle flat to convex, epaleate, but with bristles or setae; ray flowers absent; disc flowers numerous, tubular, bisexual, fertile, purple, the tube to 25 mm long, the corolla 5-lobed, 5–7 mm long; cypselae ovoid, brown, glabrous, 3.0–4.5 mm long; pappus of numerous bristles 20–30 mm long.

Common Name: Bull thistle.
Habitat: Fields, pastures, roadsides, disturbed soil.
Range: Native to Europe and Asia; adventive throughout North America.
Illinois Distribution: Common throughout Illinois.

This stout thistle is distinguished by its large purple flowering heads, its spiny winged stems, and its leaves that have spines on the upper surface.
   Cirsium vulgare flowers from June to September.

   2. **Cirsium arvense** (L.) Scop. Fl. Carn., ed. 2, 2:126. 1772.
Serratula arvensis L. Sp. Pl. 2:820. 1753.
Serratula setosa Willd. Sp. Pl. 3:1645. 1804.
Cnicus arvensis (L.) Hoffm. Deutsch. Fl., ed. 2, 1:130. 1804.
Cirsium setosum (Willd.) Bieb. Fl. Taur. Cauc. 3:560. 1819.

   Perennial herbs with creeping rootstocks; stems erect, to 90 cm tall, branched or unbranched, glabrous or gray-tomentose; basal leaves absent at flowering time; cauline leaves simple, alternate, to 25 cm long, to 5 cm wide, oblong to elliptic-lanceolate, entire or spinulose-dentate or undulate or crenate or pinnatifid, glabrous to gray-tomentose, acuminate at the apex, sessile and sometimes clasping at the base; heads numerous, in corymbs or panicles, discoid, to 2 cm across, on peduncles up to 7 cm long; involucre ovoid to campanulate, 15–20 mm high; phyllaries many, in 6–8 series, entire to erose, short spine-tipped, the outer ovate, appressed, the inner linear, usually spreading; receptacle flat or convex, epaleate but with bristles or setae; ray flowers absent; disc flowers numerous, tubular, purple or pink, rarely white, some staminate, others pistillate and fertile, the tube in the staminate flowers 12–18 mm long, the tube in the pistillate flowers 14–20 mm long, the corolla 5-lobed, the lobes 2–5 mm long; cypselae ovoid, brown, glabrous, 2–4 mm long; pappus of numerous bristles up to 30 mm long.
   There are 2 readily recognizable varieties in Illinois.
a. Stems and leaves glabrous or nearly so, green. . . . . .2a. *C. arvense* var. *arvense*
a. Stems and leaves gray-tomentose . . . . .2b. *C. arvense* var. *vestitum*

   2a. **Cirsium arvense** (L.) Scop. var. **arvense**
Cirsium arvense (L.) Scop. var. *horridum* Wimm. & Grab. Fl. Siles. 2:82. 1829.
Cirsium arvense (L.) Scop. var. *mite* Wimm. & Grab. Fl. Siles. 2:82. 1829.
Cirsium arvense (L.) Scop. var. *integrifolium* Wimm. & Grab. Fl. Silas. 2:82. 1829.
Cnicus arvensis (L.) R. Hoffm. f. *albiflorus* E. L. Rand & Redfield, Proc. Bost. Soc. Nat. Hist. 36:340. 1922.
Cirsium arvense (L.) Scop. f. *albiflora* (E. L. Rand & Redfield) Ralph Hoffm. Proc. Bost. Soc. Nat. Hist. 36:340. 1922.

   Stems and leaves glabrous or nearly so, green.

Common Name: Canada thistle; field thistle.
Habitat: Disturbed soil.
Range: Native to Europe and Asia (not Canada); adventive throughout North America.
Illinois Distribution: Common throughout the state.

This is a common, aggressive plant of roadsides and other disturbed areas. It has the smallest flowering heads of any species of *Cirsium* in Illinois. The plants are dioecious. Rare white-flowered plants may be called f. *albiflorum*.

Most plants have pinnatifid leaves with numerous spines. Those with weak spines are the typical variety. Those with stiff spines have been called var. *horridum*. Some plants have undulate leaves with weak spines. These have been called var. *mite*. Other plants have leaves nearly entire or crenate and scarcely prickly. These have been called *Cirsium setosum* or *Cirsium arvense* var. *integrifolium*. Frequently in colonies of *C. arvense* are individual plants that are yellow and etiolated.

This variety flowers from May to October.

2b. **Cirsium arvense** (L.) Scop. var. **vestitum** Wimm. & Grab. Fl. Siles. 2:92. 1829.
Stems and leaves gray-tomentose; leaves entire or undulate, scarcely and weakly prickly.

Common Name: Gray Canada thistle.
Habitat: Disturbed soil.
Range: Native to Europe and Asia; in most of the United States.
Illinois Distribution: Scattered throughout the state.

This variety flowers from May to October.

3. **Cirsium altissimum** (L.) Sprengel, Syst. Veg. 3:373. 1826.
Carduus altissimus L. Sp. Pl. 2:824. 1753.
Cnicus altissimus (L.) Willd. Sp. Pl. 3:1671. 1804.
Cirsium virginianum (L.) Willd. var. *filipendulum* Gray, Man. Bot., ed. 2, 233. 1856.
Cirsium altissimum (L.) Sprengel f. *moorei* Steyerm. Rhodora 41:585. 1939.

Biennial or perennial herbs from a taproot and fibrous roots; stems erect, branched, to 3 m tall, villous with septate hairs; basal leaves absent at flowering time; cauline leaves simple, alternate, narrowly lanceolate to oblong to ovate, to 40 cm long, the 10 (–12) cm wide, toothed or very shallowly pinnatifid, flat, spine-tipped, densely white-tomentose on the lower surface, tapering to a narrowly winged petiole; heads 1 to many, in corymbs or panicles, 3.5–4.5 mm across, discoid, not subtended by spiny bracts, on peduncles up to 5 cm long; involucre ovoid to campanulate, 2.5–3.5 cm high; phyllaries several, in 10–20 series, unequal, spine-tipped, the outer ovate, entire, appressed, the inner linear-lanceolate, erose or serrulate, spreading; receptacle flat to convex, epaleate, but with bristles or setae; ray flowers absent; disc flowers numerous, tubular, bisexual, fertile, pink or purple, rarely white, the tube up to 15 mm long, the corolla 5-lobed, 5–9 mm long;

cypselae ovoid, glabrous, 4–6 mm long, tan or brown; pappus of numerous bristles up to 20 (–30) mm long.

Common Name: Tall thistle.
Habitat: Woods, oak savannas, disturbed soil.
Range: Massachusetts to North Dakota, south to Texas and Florida.
Illinois Distribution: Occasional to common throughout the state.

This species is similar to *C. discolor*, but it has leaves that are mostly toothed or shallowly pinnatifid and cypselae that are longer. Young leaves may be nearly entire. Rare white-flowered plants may be called f. *moorei*.

Hybrids between this species and *C. discolor* have been reported.

Cirsium altissimum flowers during August and September.

4. **Cirsium discolor** (Muhl. ex Willd.) Spreng. Syst. Veg. 3:371. 1836.
Carduus discolor (Muhl. ex Willd.) Nutt. Gen. N. Am. Pl. 2:130. 1818.
Cnicus altissimus (L.) Willd. var. *discolor* (Muhl.) Gray, Proc. Am. Acad. 19:56. 1884.
Cirsium altissimum (L.) Spreng. f. *albiflorum* Britt. Bull. Torrey Club 17:125. 1890.
Cirsium discolor (Muhl. ex Willd.) Spreng. f. *albiflorum* (Britt.) House, N.Y. State Mus. Bull. 243–44:55. 1921.

Biennial or perennial herbs from a taproot and fibrous roots; stems erect, branched, to 2.5 m tall, villous with septate hairs, or sometimes nearly glabrous; basal leaves absent at flowering time, cauline leaves simple, alternate, lanceolate to ovate, to 25 cm long, to 15 cm wide, deeply pinnatifid, the lobes spine-tipped and more or less revolute, densely white-tomentose on the lower surface, tapering to the base; heads 1 to many, in corymbs or panicles, 4.0–4.5 cm across, discoid, not subtended by spiny bracts, on peduncles up to 5 cm long; involucre 2.5–3.5 cm high, ovoid to campanulate; phyllaries several in 10–12 series, unequal, spine-tipped, the outer ovate, entire, appressed, the inner lanceolate, serrulate, the apex narrowed, spreading; receptacle flat to convex, epaleate but with bristles or setae; ray flowers absent; disc flowers numerous, tubular, pink or purple, bisexual, fertile, the tube 12–15 mm long, the corolla 5-lobed, 6–9 mm long; cypselae ovoid, glabrous, 4–5 mm long; pappus of numerous bristles up to 2.5 mm long.

Common Name: Pasture thistle; field thistle.
Habitat: Fields, open woods, degraded prairies, black oak savannas, roadsides.
Range: Quebec to Saskatchewan, south to Kansas, Louisiana, and Georgia.
Illinois Distribution: Common throughout Illinois.

This species is distinguished by its deeply pinnatifid leaves that are revolute and densely white-tomentose on the lower surface.

White-flowered specimens may be called f. *albiflorum*.

Cirsium discolor flowers from August to October.

5. **Cirsium muticum** Michx. Fl. Bor. Am. 2:89. 1803.
Cnicus muticus (Michx.) Pursh, Fl. Am. Sept. 506. 1814.
Cnicus glutinosus Bigel, Fl. Bost., ed. 2, 291. 1824.
Cirsium muticum Michx. f. *lactiflorum* Fern. Rhodora 35:369. 1933.

Biennial herbs from a taproot; stems erect, branched above, hollow, villous or tomentellous, to 2.5 m tall; basal leaves, when present, deeply pinnatifid, on petioles; cauline leaves alternate, simple, deeply pinnatifid, to 50 cm long, to 20 cm wide, the lobes linear to oblòng to lanceolate, tomentose on the lower surface or sometimes nearly glabrous, spine-tipped; heads 1 to several, in corymbs or panicles, discoid, 2.5–3.5 cm across, on open peduncles up to 15 cm long, subtended by a ring of spiny bracts; involucre ovoid to cylindric, 1.5–3.0 cm high; phyllaries several in 8–12 series, unequal, not spine-tipped or with minute spines up to 0.5 mm long, viscid, the outer ovate, appressed, the inner linear-lanceolate, appressed; receptacle flat to convex, epaleate, but with bristles or setae; ray flowers absent; disc flowers numerous, tubular, rose-purple, rarely white, bisexual, fertile, the tube 10–15 mm long, the corolla 5-lobed, 4–8 mm long; cypselae ovoid, dark brown, glabrous, 4.5–5.5 mm long; pappus of numerous bristles up to 20 mm long.

Common Name: Swamp thistle; fen thistle.
Habitat: Calcareous fens.
Range: Labrador to Manitoba, south to Texas and Florida.
Illinois Distribution: Occasional in the northern one-third of Illinois, south to
    Wabash County. Surprisingly, this species has not been found in southern
    Illinois, although it occurs in states south of Illinois.

The viscid phyllaries are the most distinguishing features of this species. Because of the absence of spine-tipped phyllaries, it is the only thistle in Illinois with flowering heads that one can handle without feeling prickles. White-flowered plants may be called f. *lactiflorum*.
    This is a great indicator species for calcareous fens.
    Cirsium muticum flowers from August to October.

6. **Cirsium hillii** (Canby) Fern. Rhodora 10:95. 1908.
Cnicus hillii Canby, Gard. & For. 4:101. 1891.
Carduus hillii (Canby) Porter, Mem. Torrey Club 5:344. 1894.
Cirsium pumilum (Nutt.) Spreng. var. *hillii* (Canby) B. Boivin, Nat. Can. 94:646. 1972.

Biennial or perennial herbs from hollow, tuberous-thickened roots; stems erect, with short branches, villous to tomentose, sometimes arachnoid, to 60 cm tall; basal leaves usually present at flowering time; cauline leaves alternate, pinnately lobed or shallowly pinnatifid, the lobes obtuse with slender spine tips, to 30 cm long, to 10 cm wide, gray-hairy beneath, becoming glabrous, sometimes arachnoid, the uppermost often auriculate-clasping; heads 1 to few, in corymbs or panicles, 5–7 cm across, discoid, subtended by several bracts, on peduncles up to 15 cm long; involucre ovoid to cylindric, 3.5–5.0 cm high; phyllaries several, in 8–10 series, unequal, spine-tipped, with the spines slender, 1.5–3.0 mm long, the outer lanceolate to ovate, the inner linear-lanceolate, erose; receptacle flat or convex, epaleate, but with bristles and setae; ray flowers absent; disc flowers numerous, tubular, pink or purple, bisexual, fertile, the tube 2.5–3.5 mm long, the corolla 5-lobed, 8–10 mm long; cypselae ovoid, pale brown, glabrous 4.5–5.0 mm long; pappus of numerous bristles 4.0–4.5 mm long.

Common Name: Hill's thistle.
Habitat: Prairies, oak savannas, gravel hill prairies.
Range: Ontario to Minnesota, south to Iowa and Indiana.
Illinois Distribution: Northern three-fifths of Illinois, but becoming rare.

This species is distinguished by its stems and leaves gray-hairy to nearly glabrous, its appressed phyllaries, and its inner phyllaries erose.

Several botanists consider this plant to be a variety of *C. pumilum*. Although some intermediate specimens have been seen, mostly toward the eastern side of the range of *C. hillii*, I believe there is enough evidence to recognize *C. hillii* as a distinct species. The following table summarizes the differences:

	C. hillii	*C. pumilum*
taproot	enlarged, hollow	slender, solid
stems	with short branches	with long branches
leaves	shallowly lobed	deeply lobed
leaf lobes	obtuse, slender spine-tipped	acute, flat spine-tipped
outer phyllaries	weak spines 1.5–3.0 mm long	stout spines 4–6 mm long
corolla	4.5–5.5 mm long	4.0–4.5 mm long
cypselae	4.5–5.0 mm long	3.5–4.0 mm long
habitat	prairies, oak savannas	fields, woods

Cirsium hillii flowers from June to August.

7. **Cirsium carolinianum** (Walt.) Fern. & Schub. Rhodora 50:229. 1948.
Carduus carolinianus Walt. Fl. Carol. 195. 1788.

Biennial herbs from a slender taproot and fibrous roots; stems erect, sparsely branched, glabrous to villous, to 1.75 m tall; basal leaves present at flowering time, oblanceolate, thin, densely white-tomentose on the lower surface, on long petioles; cauline leaves alternate, simple, linear to elliptic, to 30 cm long, to 5 cm wide, dentate to shallowly lobed, the teeth and lobes spine-tipped, with the spines up to 5 mm long, sessile or on short petioles; heads1 to many, in panicles, discoid, on peduncles up to 15 cm long; involucre ovoid to campanulate, 1.2–2.0 cm high; phyllaries several, in 7–10 series, unequal, the outer linear to lanceolate, spine-tipped, with the spines 1–4 mm long, the inner linear, acuminate and twisted at the apex;

receptacle flat to convex, epaleate, but with bristles or setae; ray flowers absent; disc flowers numerous, tubular, pink or purple, bisexual, fertile, the tube up to 10 mm long, the corolla 5-lobed, the lobes 4–5 mm long; cypselae ovoid, glabrous, pale brown, 3–4 mm long; pappus of numerous bristles 12–14 mm long.

Common Name: Carolina thistle; small-headed thistle.
Habitat: Dry, open woods.
Range: Ohio to Missouri, south to Texas and Georgia.
Illinois Distribution: Extreme southeastern counties of Illinois.

This species has the smallest flowering head of any of our native species of *Cirsium*. The cypselae, which are 3–4 mm long, are shorter than those of *C. discolor* or *C. altissimum*.

Cirsium carolinianum flowers during June and July.

8. **Cirsium flodmanii** (Rydb.) Arthur, Torreya 12:34. 1912.
Carduus flodmanii Rydb. Mem. N. Y. Bot. Gard. 1:451. 1900.

Biennial herbs from shallow horizontal runners; stems erect, usually branched, densely white-tomentose, to 1 m tall; basal leaves withered at flowering time; cauline leaves alternate, simple, deeply pinnatifid or sometimes only coarsely dentate, to 40 cm long, to 10 cm wide, the tips and lobes spine-tipped with spines up to 7 mm long, the upper surface sparsely tomentose, the lower surface densely white-tomentose; heads 1 to few, in corymbs, 3.0–4.5 cm across, discoid, not subtended by spiny bracts, on peduncles up to 5 cm long; involucre campanulate, 2.0–3.5 mm high; phyllaries several, in 7–12 series, unequal, the outer lanceolate to ovate, spine-tipped, with the spines 2–4 mm long, the inner linear, not spine-tipped; receptacle flat to convex, epaleate, but with bristles or setae; ray flowers absent; disc flowers numerous, tubular, purple, bisexual, fertile, the tube up to 15 mm long, the corolla 5-lobed, the lobes up to 10 mm long; cypselae ovoid, glabrous, pale brown, 3–5 mm long; pappus of numerous bristles up to 30 mm long.

Common Name: Prairie thistle; Flodman's thistle.
Habitat: Prairies.
Range: Quebec to Alberta, south to Colorado and Illinois.
Illinois Distribution: Rare in the northern counties in the state.

This rare species of prairies is distinguished by its white-tomentose stems and its cypselae 3–5 mm long.

Cirsium flodmanii flowers from June to September.

9. **Cirsium undulatum** (Nutt.) Spreng. Syst. Veg. 3:374. 1826.
Carduus undulatus Nutt. Gen. N. Am. Pl. 2:130. 1818.

Biennial herbs from deep horizontal runners; stems erect, branched or unbranched, densely tomentose, to 1 m tall; basal leaves usually present at flowering time, on winged petioles; cauline leaves alternate, simple, to 40 cm long, to 10 cm wide, usually undulate but sometimes dentate or lobed, spine-tipped, the spines up to 10 mm long, usually revolute, sparsely tomentose on the upper surface, densely white-tomentose on the lower surface; heads 1 to few, in corymbs, discoid, not subtended by spiny bracts, on peduncles up to 25 cm long; involucre ovoid to hemispheric, 2.5–3.5 cm high; phyllaries several, in 8–12 series, unequal, the outer lanceolate to ovate, spine-tipped, the spines up to 5 mm long, the inner narrowly lanceolate, minutely spine-tipped or not spine-tipped; receptacle flat to convex, epaleate, but with bristles or setae; ray flowers absent; disc flowers numerous, tubular, pink or purple, bisexual, fertile, the tube up to 25 (–28) mm long, the corolla 5-lobed, the lobes up to 12 mm long; cypselae ovoid, glabrous, brown, 6–7 mm long; pappus of numerous bristles up to 35 mm long.

Common Name: Wavy-leaved thistle.
Habitat: Disturbed soil.
Range: Native to the western United States; adventive in Illinois.
Illinois Distribution: Known only from Will County, where it was collected by Homer Skeels in 1904 along the Rock Island Railroad in Joliet.

The densely tomentose stems and the long cypselae distinguish this species.
   Cirsium undulatum flowers during July and August.

   10. **Cirsium pitcheri** (Torr. ex Eaton) Torr. & Gray, Fl. N. Am. 2:456. 1843.
Cnicus pitcheri Torr. ex Eaton, Man. Bot., ed. 5, 180. 1829.

   Biennial or perennial herbs with an exceptionally long taproot; stems erect, sometimes branched, usually narrowly winged, densely gray-tomentose, to 85 cm tall; basal leaves present or withered at flowering time, petiolate; cauline leaves alternate, simple, elliptic to obovate, to 30 cm long, to 15 cm wide, revolute, gray-tomentose on both surfaces, decurrent at the base, spinulose, the spines 1–2 mm long; heads 1 to several, in corymbs, 3–4 cm wide, discoid, on peduncles up to 5 cm long; involucre ovoid to campanulate, 2–3 cm high; phyllaries several, in 6–8 series, unequal, the outer narrowly ovate, spine-tipped, the spines 1–2 mm long, the inner linear-lanceolate, not spine-tipped; receptacle flat to convex, epaleate, but with bristles or setae; ray flowers absent; disc flowers numerous, tubular, cream-colored, bisexual, fertile, the tube 10–15 mm long, the corolla 5-lobed, the lobes 4–8 mm long; cypselae ovoid, glabrous, pale brown, 6.0–7.5 mm long; pappus of numerous bristles up to 15 mm long.

Common Name: Dune thistle.
Habitat: Sand dunes.
Range: Ontario, Wisconsin, Michigan, Illinois, and Indiana.
Illinois Distribution: Along Lake Michigan in Cook and Lake counties.

This species used to occur in sand along Lake Michigan in both Cook and Lake counties, but it has not been found in Illinois in recent years because of destruction of shoreline habitat.
   The cream-colored flowering heads and decurrent leaves are distinctive.
   Cirsium pitcheri flowers during June and July.

92. **Arctium** L.—Burdock

Biennial or perennial herbs without spines; stems ascending to erect, branched; leaves basal and cauline, petiolate, alternate, entire to lobed, glandular-dotted; heads discoid, bracteate, in racemes, panicles, or corymbs; involucre ovoid to sub-globose; phyllaries in many series, the outer with hooked spiny tips, the inner with straight spiny tips; receptacle flat, epaleate, but with subulate scales; ray flowers absent; disc flowers 5–40, tubular, bisexual, fertile, pink or purple, the corolla 5-lobed; cypselae obovoid, flattened, usually ribbed, glabrous; pappus of numerous capillary bristles, caducous.

This genus comprises 10 species native to Europe, Asia, and North Africa.

Although the hooked spines of the outer phyllaries are similar to the spines in *Xanthium*, they do not penetrate the skin when one handles them. However, the burs stick readily to clothing and to the fur of mammals when mature, providing a good mechanism for seed dispersal.

Three introduced species occur in Illinois.

1. Heads more or less corymbose, long-pedunculate; petioles strongly angled, solid or hollow.
    2. Petioles solid; involucre glabrous or nearly so; heads 3.0–4.5 cm across . . 1. *A. lappa*
    2. Petioles hollow; involucre tomentose; heads 1.5–2.5 cm across . . . . 2. *A. tomentosum*
1. Heads more or less racemose, sessile or short-pedunculate; petioles scarcely angled, hollow . . . . 3. *A. minus*

1. **Arctium lappa** L. Sp. Pl. 2:816. 1753.
Lappa officinalis All. Fl. Ped. 1:145. 1785.
Lappa major Gaertn. Fruct. & Sem. 2:379. 1802.

Biennial herbs to 3 m tall; stems erect, much branched, pubescent; leaves basal and cauline, alternate, to 80 cm long, to 70 cm wide, obtuse to subacute at the apex, cordate or truncate at the base, entire or coarsely dentate, glabrous or sparsely pu-bescent on the upper surface, more or less gray-tomentose on the lower surface, the petioles to 35 cm long, strongly angular, solid, deeply furrowed, glabrous or pubes-cent; heads discoid, 3.0–4.5 cm across, arranged in corymbs on long peduncles; in-volucre subglobose, green, glabrous or nearly so; phyllaries in many series, glabrous or with spreading hairs, the outer linear, to 10 mm long, with hooked spiny tips, the inner linear, to 12 mm long, with straight tips; receptacle flat, epaleate, but with su-bulate scales; ray flowers absent; disc flowers about 40, tubular, bisexual, fertile, usu-ally purple, the corolla 5-lobed, 10–14 mm long, glabrous; cypselae pale brown with dark spots, 6–7 mm long, rugulose near the apex, glabrous but with spiny bristles; pappus of numerous capillary bristles up to 5 mm long, caducous.

Common Name: Great burdock.
Habitat: Disturbed soil, along railroads.
Range: Native to Europe and Asia; naturalized in much of North America.
Illinois Distribution: Occasional to common in the northern half of Illinois; also
    Clark and Hardin counties.

This species differs from the other 2 species of *Arctium* in Illinois in its solid petioles. It is a plant of disturbed soil.

Arctium lappa flowers from July to October.

2. **Arctium tomentosum** Mill. Gard. Dict., ed. 8, Arctium no. 3. 1768.

Biennial herbs to 2.5 m tall; stems erect, much branched, pubescent; leaves basal and cauline, alternate, to 40 cm long, to 30 cm wide, obtuse to subacute at the apex, cordate or truncate at the base, entire to coarsely dentate, usually arachnoid and white-tomentose on the lower surface, less pubescent and green on the upper surface, the petioles up to 15 cm long, hollow, strongly angled, furrowed, glandular-pubescent; heads discoid, 1.5–2.5 cm across, arranged in corymbs, on long peduncles; involucre subglobose, gray-tomentose; phyllaries in many series, pubescent with spreading hairs, the outer linear, to 7 mm long, with hooked spiny tips, the inner linear, to 10 mm long, with straight spiny tips; receptacle flat, epaleate, but with subacute scales; ray flowers absent; disc flowers about 30, tubular, bisexual, fertile, usually purple, the corolla 5-lobed, 10–14 mm long, glandular; cypselae gray-brown, spotted, 4.5–8.0 mm long, rugulose throughout, glabrous but with spiny bristles; pappus of numerous capillary bristles up to 3 mm long, caducous.

Common Name: Cotton burdock; woolly burdock.
Habitat: Disturbed soil.
Range: Native to Europe and Asia; naturalized mostly in the northern half of North America.
Illinois Distribution: Occasional in the northern half of the state.

This species is distinguished readily from the other species of *Arctium* in Illinois by its densely white-tomentose lower leaf surfaces.

Arctium tomentosum flowers from July to October.

3. **Arctium minus** Schk. Bot. Handb. 3:49. 1803.
Lappa minor DC. Fl. Fran. 4:77. 1805.
Arctium nemorosum Lej. in Lej. & Court. Comp. Fl. Belg. 3:129. 1836.
Arctium lappa L. var. *minus* (DC.) Gray, Syn. Fl. 1, part 2:397. 1884.
Arctium minus Schk. f. *laciniatum* Clute, Am. Bot. 15:83. 1809.

Biennial herbs to 3 m tall; stems erect, much branched, pubescent; leaves basal and cauline, alternate, to 60 cm long, to 40 cm wide, obtuse to acute at the apex, rounded or sometimes truncate at the base, entire or dentate, usually with cob-webby hairs on both surfaces, the petioles to 50 cm long, hollow, scarcely angular, usually with cobwebby hairs; heads discoid, 1.5–2.5 cm across, racemose, sessile or on short peduncles; involucre subglobose, green, glabrous or nearly so; phyllaries in several series, linear to linear-lanceolate, pubescent, usually purplish, the outer up to 10 mm long, with hooked spiny tips, the inner to 12 mm long, with short, straight spiny tips; receptacle flat, epaleate, with subulate scales; ray flowers absent; disc flowers about 30, tubular, bisexual, fertile, usually purple, the corolla

5-lobed, 6 -12 mm long, glabrous; cypselae dark brown, usually spotted, 5–6 mm long, smooth or less commonly rugulose; pappus of numerous capillary bristles 1.0–3.5 mm long, caducous.

Common Name: Common burdock.
Habitat: Disturbed soil.
Range: Native to Europe and Asia; adventive throughout the United States.
Illinois Distribution: Common throughout Illinois; probably in every county.

This common weedy species is similar to *A. tomentosum*, the other *Arctium* with hollow petioles, but differs in its green rather than white-tomentose lower leaf surface.

Arctium minus flowers from June to September.

93. **Acroptilon** Cass. in F. Cuvier—Russian Knapweed

Perennial spineless herbs; stems erect, branched; leaves basal and cauline, alternate, dentate or pinnatifid, glandular-dotted; heads discoid, in corymbs or panicles; involucre subglobose to ovoid; phyllaries in 6–8 series, scarious, the innermost bristly ciliate or bristly plumose; receptacle flat, epaleate, but bearing flattened scales; ray flowers absent; disc flowers up to 35, tubular, bisexual, fertile, blue, pink, or white, the corolla with 5 linear lobes; cypselae obovoid, more or less flattened, glabrous, the attachment scars subbasal; pappus of many unequal scales or bristles, barbed, often plumose.

The genus comprises only the following Eurasian species. *Acroptilon* is usually included within *Centaurea*, differing in the absence of sterile outer flowers and the subbasal scars on the cypselae.

1. **Acroptilon repens** (L.) A. DC. in DC. Prodr. 6:663. 1838.
Centaurea repens L. Sp. Pl., ed. 2, 2:1293. 1768.
Centaurea picris Pallas ex Willd. Sp. Pl., ed. 4, 3:2302. 1803.
Acroptilon picris (Pallas ex Willd.) C. A. Meyer, Verz. Pfl. Casp. Meer. 67. 1831.
Rhaponticum repens (L.) Hidalgo, Ann. Bot. Oxford 97:714. 2006.

Perennial herbs with creeping roots; stems erect, branched, to 60 cm tall, cobwebby-tomentose; leaves basal and cauline, the basal often absent by flowering time, the cauline alternate, to 15 cm long, to 7 cm wide, linear-lanceolate to narrowly oblong, dentate or pinnatifid, puberulent or tomentellous or glabrate, sessile or on short petioles; heads discoid, in corymbs or panicles, leafy bracted, pedunculate; involucre ovoid, 9–15 mm high, more or less cobwebby-tomentose; phyllaries in 6–8 series, the outer ovate, scarious, obtuse to acute at the apex, the inner lanceolate, acute to acuminate, sometimes bristly-plumose; receptacle flat, epaleate, but with setiform bristles; ray flowers absent; disc flowers up to 35, tubular, bisexual, fertile, blue, pink, or white, the corolla 5-lobed; cypselae obovoid, more or less flattened, light gray to brown, 2–4 mm long, glabrous; pappus of several white capillary bristles up to 10 mm long, caducous.

Common Name: Russian knapweed.

Habitat: Disturbed soil.

Range: Native to Asia; adventive in most of the United States except for the southeastern states.

Illinois Distribution: Known only from DeKalb, Williamson, and Winnebago counties.

This species is reputed to be poisonous to horses.

Acroptilon repens flowers from May to September.

94. **Amberboa** (Pers.) Less.—Sweet Sultan

Annual or biennial herbs without spines; stems erect, branched; leaves basal and cauline, alternate, entire, dentate, or lobed, glabrous or pubescent; head solitary, radiate, on a long bractless peduncle; involucre ovoid; phyllaries in several series, scarious, sometimes spine-tipped; receptacle flat, epaleate, but with setiform scales; flowers many per head, the outer sterile with raylike corollas, the inner tubular, bisexual, fertile, pink, purple, or yellow; cypselae oblongoid, more or less flattened, ribbed, with long hairs or glabrous, denticulate at the apex; pappus in several series, of distinct scales.

There are 6 species of *Amberboa*, all native to the Mediterranean region to central Asia. Only the following has been introduced into the United States.

1. **Amberboa moschata** (L.) DC. in DC. Prodr 6:566. 1838.

Centaurea moschata L. Sp. Pl. 2:909. 1753.

Annual herbs with fibrous roots; stems erect, branched, to 50 cm tall, glabrous; leaves cauline, alternate, oblanceolate, acute at the apex, tapering to the base, dentate to pinnatifid, glabrous, to 25 cm long, to 15 cm wide, sessile or on short petioles; head solitary, radiate, up to 5 cm across, on a long, bractless peduncle; involucre ovoid, sparsely pubescent; phyllaries in several series, orbicular, obtuse at the apex, with scarious margins, the inner with oblong appendages; receptacle flat, epaleate, but with setiform scales; flowers many per head, fragrant, white, pink, purple, or yellow, the outer sterile, raylike, many-lobed, the inner tubular, bisexual, fertile; cypselae oblongoid, more or less flattened, dark brown, glabrous, ribbed, denticulate at the apex, 3.5–4.0 mm long; pappus in several series, of distinct scales 3.5–4.0 mm long.

Common Name: Sweet sultan.

Habitat: Disturbed soil.

Range: Native to Asia; rarely established in the United States.

Illinois Distribution: Known only from Champaign County. The previous report from DuPage County was based on a misidentification.

This attractive species is occasionally grown in gardens but rarely escapes from cultivation.

Often included in *Centaurea*, *Amberboa* differs in its denticulate-tipped cypselae and a conspicuous rim around the basal scar of the cypselae.

Amberboa moschata flowers from July to September.

95. **Plectocephalus** D. Don in R. Sweet—Basket-flower

Annual herbs without spines; stems erect, branched; leaves basal and cauline, alternate, entire or dentate, glandular-dotted; heads radiate, borne singly or several in cymes; involucre hemispheric to campanulate; phyllaries many, in several series, unequal, linear, fringed at the apex; receptacle flat, epaleate, but with bristles; flowers many per head, the outer neutral, raylike, pink or purple, the inner tubular, bisexual, fertile, pink or purple or pale yellow; cypselae usually ovoid, more or less flattened, shallowly ribbed, with basal scars oblique; pappus of 1–3 stiff, barbed bristles, caducous.

There are 4 species in the genus, 2 of them native to the southwestern United States and Mexico, 1 native to South America, and 1 native to Africa. One introduced species has been found in Illinois.

The species of *Plectocephalus* are often included within *Centaurea* but differ in technical characters of the cypselae.

1. **Plectocephalus americanus** (Nutt.) D. Don in R. Sweet, Brit. Fl. Gard., ser. 2, 1: plate 51. 1830.
Centaurea americana Nutt. Journ. Phila. Acad. 2:117. 1821.

Annual herbs from fibrous roots; stems erect, usually branched, to 60 cm tall, scabrous, glandular-pubescent; leaves basal and cauline, alternate, the basal oblong to spatulate, obtuse to subacute at the apex, tapering to the base, 10–20 cm long, entire or dentate, sessile or on short, winged petioles, absent at flowering time, the cauline oblong to lanceolate, acute at the apex, tapering to the base, 5–10 cm long, entire to serrulate, sessile; head borne singly, radiate, the peduncle thickened near the tip; involucre hemispheric, 3.5–4.5 cm high; phyllaries many, in several series, unequal, entire, with a fringe at the apex, the outer elliptic, the inner linear, whitish, the fringe with 9–15 spinelike teeth 2–3 mm long, glabrous or cobwebby-tomentose; receptacle flat, epaleate, but with bristles; flowers many per head, the outer neutral, pink to purple, raylike, 35–50 mm long, the central flowers tubular, bisexual, fertile, pinkish, 20–25 mm long; cypselae ovoid, somewhat flattened, 4–5 mm long, gray-brown to black, glabrous or with some pubescence near the base, shallowly ribbed; pappus of several stiff, unequal bristles up to 15 mm long, caducous.

Common Name: American basket-flower.
Habitat: Disturbed soil.
Range: Native to the southwestern United States; sometimes cultivated as an ornamental but rarely escaped.
Illinois Distribution: Known only from Lawrence and Wabash counties.

Plectocephalus americanus flowers from July to September.

96. **Carthamus** L.—Safflower

Annual or perennial herbs; stems erect, branched; leaves basal and/or cauline, alternate, dentate to pinnately lobed, sessile or on winged petioles; heads discoid, borne singly or few in cymes; involucre ovoid; phyllaries many, in 4–5 series, with spiny or spine-tipped appendages; receptacle convex to conical, epaleate, but with subulate scales; flowers up to 60 per head, tubular, bisexual, fertile, yellow or red or purple, usually 5-lobed; ray flowers absent; cypselae oblongoid to obpyramidal, 4-angled, usually glabrous, with lateral scars; pappus absent or of many overlapping scales in several series.

This genus comprises 14 species, most of them native to the Mediterranean area. Only the following European species has been found in Illinois.

1. **Carthamus tinctorius** L. Sp. Pl. 2:830. 1753.

Annual herbs from fibrous roots; stems erect, branched, usually glabrous; leaves cauline, alternate, lanceolate to broadly ovate, acute at the apex, tapering to the base, to 8 cm long, to 4 cm wide, dentate, the teeth spine-tipped, lustrous, glabrous, sessile or on short winged petioles; heads discoid, usually few, borne in cymes; involucre ovoid, glabrous; phyllaries many, in 4–5 series, spreading to reflexed, the outer much longer than the inner, with spiny-toothed terminal appendages; receptacle convex to conical, epaleate, but with subulate scales; ray flowers absent; disc flowers up to 60 per head, tubular, bisexual, fertile, yellow or red, the corolla 5-lobed, 20–30 mm long; cypselae usually oblongoid, 4-angled, 7–9 mm long, white, somewhat roughened; pappus absent or of several scales 1–4 mm long.

Common Name: Safflower.
Habitat: Disturbed soil.
Range: Native to Europe; occasionally escaped from cultivation but rarely persisting in the United States.
Illinois Distribution: Known only from DuPage County.

This species is cultivated for its oily seeds that are a source of vegetable dye, for its use as bird seeds, and for its ornamental value. It rarely escapes from cultivation.

Carthamus tinctorius flowers during July and August.

97. **Centaurea** L.—Knapweed; Star Thistle

Annual, biennial, or perennial herbs; stems usually erect, branched or unbranched; leaves basal and/or cauline, alternate, often deeply lobed, spiny (in one species); heads discoid, disciform, or radiate, borne singly or in corymbs; involucre campanulate, cylindric, ovoid, or hemispheric; phyllaries many, in 6 or more series, the apex fringed or dentate and sometimes spine-tipped; receptacle flat, epaleate, but with bristles; flowers usually many per head, the outer sterile with narrow corollas, the inner bisexual, fertile, variously colored, the corolla usually 5-lobed; cypselae oblongoid to obovoid, flat or somewhat 4-angled, glabrous, ribbed, with lateral attachment scars; pappus absent or of 1–3 series of stiff bristles or scales.

There are about 500 species in the genus, all native to Europe, Asia, and North Africa.

A few species traditionally placed in *Centaurea* now have been placed in other genera, *Acroptilon*, *Amberboa*, and *Plectocephalus*. On the other hand, a plant traditionally known as *Cnicus benedictus* has been transferred to *Centaurea*.

Several introduced species and hybrids have been found in Illinois.

Centaurea is taxonomically difficult because of the variability of the species, the occurrence of hybrids, and cross-breeding for horticultural purposes.

1. Phyllaries spine-tipped.
  2. Corolla yellow.
    3. Central spine of phyllaries 10–25 mm long .14. *C. solstitialis*
    3. Central spine of phyllaries up to 10 mm long.
      4. Central spine of phyllaries divided, about 5 mm long . . . . .12. *C. benedicta*
      4. Central spine of phyllaries undivided, 5–10 mm long . . . . 13. *C. melitensis*
  2. Corolla white, pink, or purple.
    5. Central spine of phyllaries 1.5–3.0 mm long.
      6. Phyllaries pale green or whitish; pappus absent or up to 0.5 mm long. ..9. *C. diffusa*
      6. Phyllaries blackish; pappus 1–3 mm long . . . . .10. *C. X psammogena*
    5. Central spine of phyllaries 10–25 mm long . . . . 11. *C. calcitrapa*
1. Phyllaries not spine-tipped.
  7. Plants annual; cypselae 4–5 mm long . . . . .1. *C. cyanus*
  7. Plants perennial; cypselae up to 4 mm long, or 5–6 mm long.
    8. Outer flowers with the corolla 25–45 mm long; cypselae 5–6 mm long . . . .
    . . . . .2. *C. montana*
    8. Outer flowers with the corolla 15–18 (–25) mm long; cypselae 2.5–3.5 mm long.
      9. Inner phyllaries pectinately fringed along the margins.
        10. Corolla up to 25 mm long; cypselae 3.0–3.5 mm long. . . . . 8. *C. stoebe*
        10. Corolla 15–18 mm long; cypselae 2.5–3.0 mm long.
          11. Heads with only disc flowers; pappus black . . . . .4. *C. nigra*
          11. Heads with ray flowers; pappus not black.
            12. Involucre as wide as high; phyllaries brown. . . . . 5. *C. X monctonii*
            12. Involucre longer than wide; phyllaries blackish . . . . . .6. *C. nigrescens*
      9. Inner phyllaries dentate or shallowly lobed along the margins.
        13. Corolla 15–18 mm long; cypselae 2.5–3.0 mm long . . . . 3. *C. jacea*
        13. Corolla 20–25 mm long; cypselae 3–4 mm long . . . . 7. *C. phrygia*

1. **Centaurea cyanus** L. Sp. Pl. 2:911. 1753.

Annual herbs from fibrous roots; stems ascending, branched, to 1 m tall, tomentose, at least when young; leaves basal and cauline, alternate, linear to linear-lanceolate, acuminate at the apex, tapering to the usually sessile base, to 10 cm long, to 2 cm wide, entire or less commonly lobed, often pubescent when young; heads 1 to few, in cymes, radiate, pedunculate; involucre campanulate or ovoid; phyllaries in 4 series, unequal, the outer ovate, the inner oblong, tomentose but nearly glabrous at flowering time, the margin of the upper half jagged with teeth about 1 mm long; receptacle flat, epaleate, but with bristles; flowers up to 35 per head, blue, purple, pink, or white, the outer sterile with raylike corolla lobes, 20–25 mm long, the inner tubular, bisexual, fertile, 10–15 mm long; cypselae oblongoid, more or less flat or 4-angled, 4–5 mm long, stramineous, pubescent; pappus of many stiff unequal bristles up to 4 mm long.

Common Name: Bachelor's-buttons; cornflower.
Habitat: Disturbed soil; old fields.
Range: Native to Europe; adventive in most of North America.
Illinois Distribution: Occasional throughout the state.

This is a common garden ornamental that often escapes from cultivation.

Centaurea cyanus flowers from July to September.

2. **Centaurea montana** L. Sp. Pl. 2:911. 1753.

Perennial herbs with stolons and rhizomes; stems erect, branched or un-branched, to 75 cm tall, villous or arachnoid-tomentose; leaves mostly cauline, alternate, lanceolate to ovate, acute at the apex, tapering to the usually decurrent base, to 30 cm long, to 8 cm wide, villous or tomentose but becoming more or less glabrous, entire to dentate to pinnately lobed, sessile or on short, winged petioles; heads 1 to several, in a corymb, radiate, pedunculate; involucre ovoid or cam-panulate; phyllaries many, in 6 or more series, ovate or lanceolate, with scarious margins, puberulent, with a pectinate fringe; receptacle flat, epaleate, but with bristles; flowers up to 60 per head, the outer ones sterile, 25–45 mm long, raylike, blue, with an elongated corolla tube, the disc flowers tubular, bisexual, fertile, purple, the corolla up to 20 mm long; cypselae oblongoid, brown, sericeous, 5–6 mm long; pappus of numerous bristles up to 1.5 mm long.

Common Name: Mountain cornflower.
Habitat: Disturbed soil.
Range: Native to Europe; adventive in the northern half of the United States.
Illinois Distribution: Known only from Cook County.

This species is distinguished by its outer sterile flowers of the head being 25–45 mm long, longer than in any other species of *Centaurea* in Illinois. This species is sometimes grown as a garden ornamental but rarely escapes into the wild.

The only Illinois collection was made by Duesner in 1908 in Maplewood, Cook County.

Centaurea montana flowers from July to September.

3. **Centaurea jacea** L. Sp. Pl. 2:914. 1753.
Jacea pratensis Lam. Fl. Franc. 2:54. 1779.

Perennial herbs from stolons and rhizomes; stems erect, branched or un-branched, to 1.2 m tall, villous to tomentose, usually becoming glabrous or nearly so at maturity; leaves basal and cauline, the basal elliptic to oblanceolate, acute at the apex, tapering to the petiolate base, to 25 cm long, to 10 cm wide, entire or dentate or pinnately lobed, usually pubescent when young, the cauline alter-nate, linear to lanceolate, acute at the apex, tapering to the sessile base, entire or denticulate; heads few, in a corymb, radiate, on bracteate peduncles; involucre ovoid to hemispheric; phyllaries many, in about 6 series, overlapping, lanceolate to ovate, glabrous or pubescent, usually hidden by fimbriate or lacerate appendages, membranous, pale brown; receptacle flat, epaleate, but with bristles; flowers many per head, purple, the outer sterile and longer than the inner flowers, the inner tubular, bisexual, fertile, 15–18 mm long; cypselae oblongoid, somewhat 4-angled, tan or pale brown, pubescent, 2.5–3.0 mm long; pappus absent.

Common Name: Brown knapweed.
Habitat: Disturbed soil.
Range: Native to Europe and Asia; occasionally adventive in the United States.
Illinois Distribution: Occasional in the northern half of Illinois.

This species is similar to *C. phrygia* but has smaller corollas and smaller cypselae.

Centaurea jacea is in an extremely variable complex. Some botanists, such as I, recognize several species in the complex, while other botanists consider only one variable species.

This species flowers from July to October.

4. **Centaurea nigra** L. Sp. Pl. 2:911. 1755.
Centaurea nemoralis Jordan, Pugill. Pl. Af. Bor. Hispan. 104. 1852.

Perennial herbs from stolons and rhizomes; stems erect, branched, to 1.2 m tall, villous or tomentose, scabrous; leaves basal and cauline, the basal and lower alternate cauline leaves elliptic to oblanceolate, acute at the apex, tapering to the petiolate base, to 25 cm long, to 10 cm wide, entire to dentate to shallowly lobed, scabrous, the middle and upper leaves alternate, linear to lanceolate, acute at the apex, tapering to the sessile base, entire or denticulate; heads few, in corymbs, discoid, on bracteate peduncles; involucre subglobose; phyllaries many, in 6 series, lanceolate to ovate, glabrous or pubescent, the bases overlapping, black or dark brown, with fimbriate appendages; receptacle flat, epaleate, but with bristles; disc flowers many per head, tubular, bisexual, fertile, purple, the corolla 20–25 mm long; cypselae oblongoid, more or less 4-angled, tan, finely pubescent, 3–4 mm long; pappus of numerous black, unequal bristles up to 1 mm long.

Common Name: Black knapweed.
Habitat: Disturbed soil.
Range: Native to Europe; occasionally adventive in the United States.
Illinois Distribution: Known from the northern one-sixth of Illinois.

This is the only *Centaurea* in Illinois with only discoid heads, purple flowers, fimbriate phyllaries, and black pappus.

Centaurea nigra flowers from July to September.

5. **Centaurea X monctonii** C. E. Britt. Bot. Soc. Exch. Club Brit. Isles 6:172. 1921.
Centaurea debeauxii Godron & Gren. ssp. *thuillieri* Dostal, Bot. Journ. Linn. Soc.
    71:207. 1976.
Centaurea nigra L. var. *radiata* DC. Cat. Pl. Hort. Monsp. 91. 1813.
Centaurea thuillieri (Dostal) J. Duvigne & Lambinon, Nouv. Fl. Belgique,
    Luxembourg, N. France, ed. 2, 829. 1978.

Perennial herbs from stolons and rhizomes; stems ascending to erect, branched, to 1.5 cm tall, villous or tomentose, scabrous, but usually nearly glabrous at maturity; leaves basal and cauline, the basal and lowermost alternate cauline leaves elliptic to lanceolate, acute at the apex, tapering to the petiolate base, to 25 cm long, to 10 cm wide, entire to dentate to pinnately lobed, pubescent, the middle and upper leaves alternate, linear to lanceolate, tapering to the sessile base, entire to dentate, pubescent; heads few, in a corymb, radiate, on bracteate peduncles; involucre ovoid to hemispheric, as wide as high; phyllaries many, in 6 series, lanceolate to ovate, pubescent or glabrous, brown, overlapping, the margins dentate to pectinately dissected; receptacle flat, epaleate, but with bristles; flowers many per head, purple, the outermost sterile, the inner tubular, bisexual, fertile, with the corolla 15–18 mm long; cypselae oblongoid, tan, pubescent, 2.5–3.0 mm long; pappus absent or of many unequal bristles up to 1 mm long.

Common Name: Meadow knapweed.
Habitat: Disturbed soil.
Range: Probably native to Europe, but a hybrid developed in the horticultural
    industry, apparently from a cross between *C. jacea* and *C. nigra*.
Illinois Distribution: Rare in Illinois (personal correspondence from David J. Kiel in
    2011).

This hybrid flowers from July to September.

   6. **Centaurea nigrescens** Willd. Sp. Pl. 3:2288. 1803.
Centaurea transalpina Schleicher ex DC. Cat. Pl. Helv., ed. 3, 11. 1815.
Centaurea vochinensis Bernh. ex Reichenb. Icon. Fl. Germ. 15:15. 1853.
Centaurea dubia Suter ssp. *nigrescens* (Willd.) Hayek, Verh. Zool.-Bot. Ges. Wien. 68:
    210. 1917.

   Perennial herbs from stolons and rhizomes; stems ascending to erect, branched, to 1.2 m tall, villous or tomentose, becoming nearly glabrous at maturity; leaves basal and cauline, the basal and lower alternate cauline leaves elliptic to oblong, acute at the apex, tapering to the petiolate base, to 25 cm long, to 10 cm wide, entire to dentate to pinnately lobed, pubescent, the middle and upper leaves alternate, linear to lanceolate, obtuse at the apex, tapering to the sessile base, entire or dentate, usually pubescent; heads few, in a corymb, radiate or discoid, on bracteate peduncles; involucre subglobose to campanulate, longer than wide; phyllaries many, in 6 series, lanceolate to ovate, pubescent or glabrous, overlapping, black, pectinately 6- to 8-lobed; receptacle flat, epaleate, but with bristles; flowers many per head, all fertile or the outermost sterile, purple, the corolla 15–18 mm long; cypselae oblongoid, tan, pubescent, 2.5–3.0 mm long; pappus absent or of several brown bristles up to 1 mm long.

Common Name: Tyrolean knapweed; Vochin knapweed.
Habitat: Disturbed soil.
Range: Native to Europe; occasionally adventive in the United States.
Illinois Distribution: Scattered in Illinois.

This species has sometimes been called *C. vochinensis* in the past in Illinois. It differs from *C. X monctonii* in that its involucre is longer than wide and it has black phyllaries. Some botanists have called this species *C. dubia*, but that is an invalid binomial.

Some flowering heads have sterile outer flowers, while others, even on the same plant, lack the sterile outer flowers.

Centaurea nigrescens flowers from July to September.

7. **Centaurea phrygia** L. Sp. Pl. 2:910. 1753.
Centaurea austriaca Willd. Sp. Pl. 3:2283. 1803.

Perennial herbs with stolons and rhizomes; stem erect, branched or un-branched, to 75 cm tall, usually pubescent; leaves basal and cauline, the basal and lower alternate cauline leaves lanceolate to ovate, acute at the apex, tapering to the winged petiole, to 15 cm long, to 12 cm wide, entire or dentate, usually pubescent, the middle and upper leaves alternate, lanceolate, sessile and sometimes clasping, usually entire; head usually solitary, radiate, on a bracteate peduncle; involucre ovoid or subglobose; phyllaries many, in 6 series, lanceolate to ovate, tomentose or glabrous, brown or black, the appendages with filiform divisions, the tips usually recurved; receptacle flat, epaleate, but with bristles; flowers many per head, pink or purple, the outer sterile and raylike, the inner tubular, bisexual, fertile, the corolla 20–25 mm long; cypselae oblongoid, tan, pubescent, 3–4 mm long; pappus absent, or of many unequal bristles up to 2 mm long.

Common Name: Wig knapweed.
Habitat: Disturbed soil.
Range: Native to Europe; adventive in a few of the eastern United States.
Illinois Distribution: Known only from DuPage County.

One of the best distinguishing characteristics of this species is the recurved tips of the phyllaries.

Centaurea phrygia flowers from July to September.

8. **Centaurea stoebe** L. ssp. **micranthos** (S. G. Gmel. ex Gugler) Hayek, Rep. Spec. Nov. Regni Veg. Beih. 30:766. 1931.
Centaurea maculosa Lam. Encycl. 1:669. 1785.
Centaurea biebersteinii DC. Prodr. 6:583. 1838.
Centaurea maculosa Lam. ssp. *micranthos* S. G. Gmel. ex Gugler, Ann. Hist.-Nat. Mus. Natl. Hung. 6:167. 1908.

Perennials with stolons and rhizomes; stems ascending to erect, branched, to 1.2 m tall, cobwebby-pubescent, becoming nearly glabrous at maturity; leaves basal and cauline, glandular-dotted, the basal and lower alternate cauline leaves deeply pinnatifid, acute at the apex, tapering to a petiolate base, gray-tomentose, becoming nearly glabrous at maturity, to 15 cm long, to 10 cm wide, the lobes linear to oblong, entire, the middle and upper leaves alternate, entire or sometimes shallowly pinnatifid, sessile; heads 1 to few, in a cyme, radiate, on bracteate peduncles; involucre ovoid; phyllaries many, in 6 series, pale green to pinkish, with dark brown or black margins, glabrous or pubescent, the outer ovate, the inner oblong, with pectinate appendages near the apex; receptacle flat, epaleate, but with bristles; flowers up to 40 per head, pink or purple, the outer sterile and raylike, the inner tubular, bisexual, fertile, with the corolla 12–15 mm long; cypselae oblongoid, pale brown to nearly white, pubescent, 3.0–3.5 mm long; pappus of many white, stiff bristles up to 5 mm long, in 1–2 series.

Common Name: Spotted knapweed.
Habitat: Disturbed soil, old fields, along railroads, pastures.
Range: Native to Europe; adventive in most of North America.
Illinois Distribution: Occasional to common in the northern half of Illinois, less
    common elsewhere.

Although this plant has been known in Illinois as *C. maculosa* or *C. biebersteinii*, it appears to be better placed in the *C. stoebe* complex. Our plant is var. *micranthos* and is a perennial, whereas typical *C. stoebe* is a biennial.

    This plant is aggressive, quickly occupying disturbed areas. It flowers from July to October.

    9. **Centaurea diffusa** Lam. Encycl. 1:675. 1785.
Acosta diffusa (Lam.) Sojak, Cas. Nar. Muz. Praze Rada Prir. 140. 1972.

    Annual or perennial herbs, usually with fibrous roots; stems ascending to erect, much branched, to 75 cm tall, gray-tomentose; leaves basal and cauline, the basal and lower cauline leaves usually absent at flowering time, to 20 cm long, to 10 cm wide, bipinnately divided, the lobes linear-oblong, hispidulous or tomentose, the middle and upper alternate, bipinnately divided or sometimes entire, sessile; heads several in panicles or corymbs, disciform, on bracteate peduncles; involucre cylindric to ovoid, much longer than wide; phyllaries many, in 6 series, pale green or whitish, the outer lanceolate to ovate, glabrous or tomentose, the margin fringed with slender spines, the spine 1.5–3.0 mm long, the inner lanceolate, fringed or spine-tipped; receptacle flat, epaleate, but with bristles; flowers up to 35 per head, white or pink, the outer sterile, 12–13 mm long, the inner tubular, bisexual, perfect; cypselae oblongoid, dark brown, usually pubescent, 3–4 mm long; pappus absent, or of bristles up to 0.5 mm long.

Common Name: Spreading star thistle.
Habitat: Disturbed soil.
Range: Native to Europe; introduced and adventive in much of the United States.
Illinois Distribution: Occasional in the northern half of Illinois.

This species differs from the other species of *Centaurea* in Illinois in the combination of spine-tipped phyllaries, white or pink corollas, and the spine of each phyllary being only up to 3 mm long.

    Centaurea diffusa flowers from July to September.

10. **Centaurea X psammogena** Gayer, publication data unclear.

Perennial herbs with fibrous roots; stems ascending to erect, branched, to 80 cm tall, tomentose; leaves basal and cauline, the basal absent at flowering time, to 20 cm long, to 10 cm wide, bipinnately divided, the lobes linear-oblong, tomentose, the middle and upper alternate, bipinnately divided, sessile; heads several in corymbs, disciform, on bracteate peduncles; involucre cylindric, longer than wide; phyllaries many, in 6 series, blackish at least in the upper half, the outer lanceolate to ovate, tomentose, the margin fringed with slender spines, the spines 1.5–3.0 mm long, the inner lanceolate, fringed or spine-tipped; receptacle flat, epaleate, but with bristles; flowers up to 35 per head, rose-colored, the outer sterile, 12–15 mm long, the inner tubular, bisexual, perfect; cypselae oblongoid, dark brown, pubescent, 3–5 mm long; pappus of short bristles up to 0.5 mm long, blackish.

Common Name: Brown hybrid knapweed.
Habitat: Disturbed soil.
Range: Native of Europe; rarely adventive in the United States.
Illinois Distribution: Known from northeastern Illinois.

This is the presumed hybrid between *C. diffusa* and *C. jacea*. It appears to be closer to *C. diffusa* but differs in its blackish-tipped phyllaries and its blackish pappus.

This hybrid flowers from June to September.

11. **Centaurea calcitrapa** L. Sp. Pl. 2:917. 1753.

Annual or rarely perennial herbs with fibrous roots; stems ascending to erect, branched, to 60 cm tall, pubescent; leaves basal and cauline, glandular-dotted, the basal and lower alternate cauline leaves pinnately divided, with narrowly lanceolate, dentate segments, to 20 cm long, to 15 cm wide, petiolate, the middle and upper leaves alternate, ovate, pinnately lobed, sessile, sometimes slightly clasping; heads 1 to several in corymbs, disciform, on bracteate peduncles; involucre ovoid, longer than wide; phyllaries many, in 6 series, the outer ovate, greenish, with a scarious margin, the appendages spiny-fringed at the base and with a stout yellow terminal spine 10–25 mm long, the inner spineless; receptacle flat, epaleate, but with bristles; cypselae oblongoid, 4-angled or sometimes flattened, white, usually with brown streaks, 2.5–3.5 mm long, usually somewhat pubescent; pappus absent.

Common Name: Purple star thistle.
Habitat: Disturbed soil.
Range: Native to Europe and North Africa; introduced and adventive in several U.S. states.
Illinois Distribution: Known only from Grundy County. It has not been seen in Illinois since 1914, when it was found by Allison near Gardner.

The purple, disciform flowering heads and the long, stout terminal spine of the outer phyllaries are distinctive for this species.

Centaurea calcitrapa flowers from June to October.

12. **Centaurea benedicta** (L.) L. Sp. Pl., ed. 2, 2:1296. 1763.
Cnicus benedictus L. Sp. Pl. 2:826. 1753.

Annual herbs with fibrous roots; stems prostrate to ascending, much branched, reddish, to 75 cm long, more or less tomentose; leaves cauline, alternate, lanceolate to oblong-lanceolate, acute at the apex, to 25 cm long, to 12.5 cm wide, dentate to pinnately lobed, the teeth and lobes with weak spines, pubescent, glandular-dotted, tapering to a short, winged petiole or sessile; head borne singly, disciform, on peduncles with bracts with spine-tipped teeth or lobes; involucre globose; phyllaries many, in 6 series, overlapping, the outer ovate, with spreading spiny tips, the inner lanceolate, with pinnately divided spines at the apex, the spines up to 5 mm long; receptacle flat, epaleate, but with long, soft hairs; flowering heads many, yellow, the outer sterile, 3-lobed, the inner tubular, bisexual, fertile, the corolla up to 25 mm long; cypselae terete, curved, 8–10 mm long, with 20 conspicuous ribs and with 10 teeth at the apex; pappus of 2 series of awns, the outer 9–10 mm long, the inner 2–5 mm long, with short, spreading hairs.

Common Name: Blessed thistle.
Habitat: Disturbed soil.
Range: Native to Europe and Asia; introduced into several states where it has escaped from cultivation as a medicinal herb.
Illinois Distribution: Known only from Champaign County.

This species is sometimes treated as the only member of the genus *Cnicus*. Linnaeus first described it as a species of *Cnicus*, later transferring it to *Centaurea*. The pappus is quite different from the pappus of the other species of *Centaurea* in Illinois, and the bracts on the peduncles have spiny teeth or lobes.

Centaurea benedicta flowers from May to September.

13. **Centaurea melitensis** L. Sp. Pl. 2:917. 1753.
Annual herbs with fibrous roots; stems erect, much branched, narrowly winged at the base of the decurrent leaves, to 1 m tall, tomentose or villous, usually scabrous; leaves basal and cauline, the basal and lower alternate cauline leaves often absent at flowering time, oblong to lanceolate, acute at the apex, tapering to the petiolate base, to 15 cm long, to 8 cm wide, entire or dentate or pinnately lobed, the middle and upper leaves alternate, linear to oblong, entire or dentate, decurrent at the base; heads 1 to few in a corymb, disciform, sessile or on short peduncles that are bracteate; involucre ovoid; phyllaries several, in 6 series, ovate, stramineous with purple appendages with a spiny-fringed base, and terminated by a slender, undivided spine 5–10 mm long, the inner entire, acute or spine-tipped; receptacle flat, epaleate, but with bristles; flowers many, in corymbs, yellow, the outer sterile, 10–12 mm long, the inner bisexual, fertile, 10–12 mm long; cypselae oblongoid, white or pale brown, pubescent, 2.0–2.5 mm long; pappus of numerous white, unequal, stiff bristles to 3 mm long.

Common Name: Maltese star thistle.
Habitat: Disturbed soil.
Range: Native to Europe, Asia, and North Africa; adventive as an escape from
    cultivation in several states.
Illinois Distribution: Known only from Menard County.

This is the only *Centaurea* in Illinois with yellow flowering heads and phyllaries
with an undivided terminal spine 5–10 mm long.

    Centaurea melitensis flowers from July to September.

    14. **Centaurea solstitialis** L. Sp. Pl. 2:917. 1753.

    Annual herbs with fibrous roots; stems erect, branched or unbranched, to 1 m
tall, tomentose; leaves basal and cauline, the basal and lower alternate cauline
leaves usually absent at flowering time, lyrate-pinnatifid, to 15 cm long, to 7 cm
wide, tomentose, scabrous, tapering to a petiole, the middle and upper leaves alternate, linear to oblong, acute or obtuse at the apex, decurrent at the base forming
a wing, entire; heads 1 to few, in corymbs, disciform, on long bracteate peduncles;
involucre ovoid or globose; phyllaries many, in 6 series, pale green, the outer
ovate, with pale brown to brown appendages with palmately radiating spines and
with a stiff yellow terminal spine up to 25 mm long, the inner linear-oblong, obtuse at the apex or with a short spine at the apex; receptacle flat, paleate, but with
a few bristles; flowers many per head, yellow, the outer sterile, 10–20 mm long, the
inner tubular, bisexual, fertile; cypselae of two types, the outer dark brown, 2–3
mm long, the inner white to pale brown, mottled, 2–3 mm long; pappus absent
from the outer cypselae and of many white, unequal bristles to 4 mm long from
the inner cypselae.

Common Name: Yellow star thistle; Barneby's thistle.
Habitat: Disturbed soil.
Range: Native to Europe; adventive in most of the United States.
Illinois Distribution: Known only from Jackson County.

This species is distinguished by its yellow flowering heads and its long central yellow spine on the outer phyllaries.

    This is an aggressive weed in the western United States. It is poisonous if eaten
by horses.

    Centaurea solstitialis flowers from July to September.

Tribe Vernonieae Cass.

Annual, biennial, or perennial herbs, sometimes woody (but not in Illinois); leaves
cauline, alternate, or sometimes both basal and cauline; heads discoid, borne
in corymbs or panicles or borne singly; involucre campanulate, hemispheric, or
cylindric; phyllaries in 2–8 series, unequal, the margins usually scarious; receptacle flat or convex, epaleate, but sometimes with bristles; ray flowers absent; disc

flowers tubular, bisexual, fertile, white, blue, pink, or purple; cypselae columnar to fusiform, sometimes flattened, sometimes ribbed; pappus in 2 series, the outer of short, stout bristles or scales, the inner of long, barbellate bristles, or less commonly the pappus in 1 series.

There may be as many as 140 genera in this tribe worldwide, with about 1,300 species. Two genera occur in Illinois.

1. Flowers up to 4 per head, the heads sessile and subtended by 3 deltate bracts . . . .
. . . . 98. *Elephantopus*
1. Flowers numerous per head, the heads pedunculate and not subtended by foliaceous
bracts . . . . 99. *Vernonia*

98. **Elephantopus** L.—Elephant's-foot

Perennial herbs from stolons and rhizomes; leaves basal and cauline, the cauline alternate, usually with winged petioles, toothed, glandular-dotted; heads discoid, in groups of up to 40 in corymbs or panicles, subtended by 2–3 deltate scales; involucre cylindric; phyllaries 8, in 4 pairs, entire, with an apiculate to spinose apex; receptacle flat, epaleate; flowers up to 4 per head, tubular, purple, the corolla with 5 unequal lobes; cypselae clavate, more or less flattened, 10-ribbed, pubescent; pappus of 5 aristate scales, persistent on the cypselae.

Elephantopus consists of about 15 species in warm temperate regions of the world, including several in the United States.

Only the following species occurs in Illinois.

1. **Elephantopus carolinianus** Raeuschel, Nom. Bot., ed. 3, 256. 1797.
Perennial herbs from stolons and rhizomes; stems erect, branched, to 80 cm tall, hirsute below, strigose above; leaves basal and cauline, the basal usually absent at flowering time, the cauline alternate, oval to ovate, obtuse to acute at the apex, tapering to the short-petiolate or sessile base, to 15 cm long, to 7.5 cm wide, crenate, pilose or hirsute on both surfaces; heads discoid, each with usually 4 flowers, subtended by 3 foliaceous, deltate bracts, the bracts to 15 mm long, to 12 mm wide; involucre cylindric; phyllaries 8, in 4 decussate pairs, 8–10 mm long, the outer 4 ovate, the inner 4 lanceolate, with a slender awn tip, pubescent; receptacle flat, epaleate; ray flowers absent; disc flowers tubular, bisexual, fertile, blue or purple; cypselae clavate, 10-ribbed, pubescent, 2.5–4.0 mm long; pappus of 5 persistent, aristate scales 4–5 mm long.

Common Name: Carolina elephant's-foot.
Habitat: Dry or mesic woods.
Range: Pennsylvania to Illinois to Kansas, south to Texas and Florida.
Illinois Distribution: Occasional in the southern one-third of Illinois.

This woodland species is readily distinguished by its blue or purple flowers that are 4 in a group, subtended by 3 foliaceous deltate bracts.

Elephantopus carolinianus flowers from July to September.

99. **Vernonia** Schreb.—Ironweed

Perennial herbs with rhizomes; stems usually stout, branched; leaves mostly cauline (in Illinois), alternate, toothed, usually glandular-dotted; heads discoid, usually purple, not subtended by foliaceous bracts; involucre campanulate to hemispheric; phyllaries many, in up to 7 series, chartaceous, entire; receptacle flat, epaleate; ray flowers absent; disc flowers tubular, bisexual, fertile, the corolla 5-lobed; cypselae cylindrical, 8- or 10-ribbed; pappus of up to 30 outer scales or bristles and up to 40 inner scales or bristles, persistent.

Vernonia consists of approximately 20 species, mostly in North America. Most of the species in the United States hybridize with one another. Five species occur in Illinois.

1. Tips of phyllaries long-filiform; heads 12–20 mm across. . . . . 1. *V. arkansana*
1. Tips of phyllaries obtuse to mucronate to abruptly acuminate; heads 4–10 mm across.
    2. Leaves tomentose to tomentellous on the lower surface.
        3. Tips of phyllaries obtuse to mucronate, erect or slightly spreading . . . 4. *V. missurica*
        3. Tips of phyllaries abruptly acuminate, recurving or more or less appressed . . . . . .
        . . . . 5. *V. baldwinii*
    2. Leaves glabrous or scabrous-hirtellous on the lower surface (sometimes tomentellous on the veins).
        4. Leaves glabrous, conspicuously punctate beneath; outer pappus of short, capillary bristles. . . . . 2. *V. fasciculata*
        4. Leaves scabrous-hirtellous, scarcely punctate beneath; outer pappus of scalelike bristles . . . . 3. *V. gigantea*

1. **Vernonia arkansana** DC. Prodr. 7:264. 1838 (April).
Vernonia crinita Raf. New Fl. N. Am. 4:77. 1838 (October).

Perennial herbs with rhizomes; stems erect, mostly unbranched, to 2.5 m tall, pubescent or nearly glabrous at maturity, sometimes glaucous; leaves cauline, alternate, linear to linear-lanceolate, acute to acuminate at the apex, tapering to the sessile base, to 15 cm long, to 2 cm wide, entire or denticulate, scabrous, glandular-dotted; heads several, in corymbs, discoid, 12–20 mm across, on peduncles to 5 cm long; involucre hemispheric; phyllaries many, in 5–6 series, scabrous, glandular-dotted, squarrose, ciliolate, the outer lance-ovate, to 10 mm long, the inner lanceolate, to 15 mm long, all with prolonged, filiform tips; receptacle flat, epaleate; ray flowers absent; disc flowers many, tubular, bisexual, fertile, purple, the corolla 5-lobed, 7–10 mm long; cypselae cylindrical, 8- or 10-ribbed, 3–5 mm long, glabrous or hispidulous on the ribs; pappus in 2 series, purplish, the outer of 25–30 scales about 1 mm long, the inner of 25–30 bristles up to 7 mm long.

Common Name: Ozark ironweed.
Habitat: Low open woods, prairies.
Range: Illinois, Missouri, Kansas, Oklahoma, and Arkansas.
Illinois Distribution: Known only from Champaign County. The original population has been destroyed.

This species is readily distinguished by the prolonged tip of the phyllaries and by its larger flowering heads 12–20 mm across.

DeCandolle and Rafinesque both described this species in 1838. DeCandolle's *V. arkansana* was published in April; Rafinesque's *V. crinita* was published in October.

Vernonia arkansana flowers during August and September.

2. **Vernonia fasciculata** Michx. Fl. Bor. Am. 2:94. 1803.

Perennial herbs with rhizomes; stems erect, branched, to 2 m tall, glabrous or less commonly finely pubescent; leaves cauline, alternate, linear-oblong to lanceolate, acuminate at the apex, tapering to the short-petiolate or sessile base, to 10 cm long, to 4 cm wide, serrate, glabrous or nearly so on both surfaces, conspicuously punctate beneath; heads many, in corymbs, discoid, on peduncles up to 1 cm long or absent; involucre campanulate; phyllaries many, in 4–5 series, the margins ciliate, the outer ovate, acute or apiculate at the apex, up to 3 mm long, the inner narrowly oblong, 5–7 mm long; receptacle flat, epaleate; ray flowers absent; disc flowers up to 30 per head, tubular, bisexual, fertile, purple, the corolla 5-lobed; cypselae cylindrical, 8- or 10-ribbed, glabrous or pubescent, 3.5–4.0 mm long; pappus purplish, in 2 series, the outer of 20–30 short, capillary bristles up to 3 mm long, the inner of up to 50 subulate scales or bristles up to 8 mm long.

Common Name: Common ironweed.
Habitat: Wet prairies, moist soil.
Range: Wisconsin to Manitoba, south to Colorado, Oklahoma, and Kentucky.
Illinois Distribution: Common in the northern half of Illinois, occasional in the southern half.

This is the only species of *Vernonia* in Illinois with acute to apiculate phyllaries and with stems and leaves glabrous or nearly so.

Vernonia fasciculata flowers from July to October.

3. **Vernonia gigantea** (Walt.) Branner & Coville, Rep. Ark. Geol. Surv. 4:189. 1891. *Chrysocoma gigantea* Walt. Fl. Carol. 196. 1788.

Perennial herbs with rhizomes; stems erect, branched, to 2 m tall, glabrous or hirtellous; leaves cauline, alternate, lanceolate to lance-ovate, acuminate at the apex, tapering to the short-petiolate or sessile base, to 25 cm long, to 5.5 cm wide, serrate, glabrous on the upper surface, hirtellous or tomentellous on the lower surface, scarcely punctate beneath; heads numerous, in corymbs, discoid, on glabrous or hirtellous peduncles up to 1.5 cm long or sessile; involucre campanulate to hemispheric; phyllaries many, in 4–5 series, ciliolate, obtuse or apiculate at the apex, the outer broadly lanceolate, 1–2 mm long, the inner oblong, 3.5–5.0 mm long; receptacle flat, epaleate; ray flowers absent; disc flowers up to 30 per head, tubular, bisexual, fertile, purple, the corolla 5-lobed; cypselae cylindric, 2.5–3.5 mm long, 8- or 10-ribbed, the ribs hirtellous; pappus in 2 series, purplish, the outer of 20–25 scales up to 1 mm long, the inner of 35–40 capillary bristles up to 6 mm long.

Two varieties occur in Illinois.

a. Lower surface of leaves glabrous or with appressed hairs. . . 3a. *V. gigantea* var. *gigantea*
a. Lower surface of leaves with spreading hairs, hirtellous or tomentellous. . . . .
. . . . .3b. *V. gigantea* var. *taeniotricha*

3a. **Vernonia gigantea** (Walt.) Branner & Coville var. **gigantea**
Chrysocoma gigantea Walt. Fl. Carol. 196. 1788.
Vernonia altissima Nutt. Gen. 2:134. 1818.

Lower surface of leaves glabrous or with appressed hairs.

Common Name: Tall ironweed.
Habitat: Low woods, open wet ground.
Range: Pennsylvania to Michigan to Nebraska, south to Texas and Florida; Ontario.
Illinois Distribution: Occasional in the southern half of Illinois, less common
    elsewhere.

This typical variety is not as common or as widespread as var. *taeniotricha*.
    For many years, this plant was known in Illinois as *V. altissima*.
    Variety *gigantea* flowers from July to October.

3b. **Vernonia gigantea** (Walt.) Branner & Coville var. **taeniotricha** S. F. Blake,
    Rhodora 19:167–68. 1917.
Vernonia X illinoensis Gl. Bull. N. Y. Bot. Gard. 4:211. 1906.

Lower surface of leaves with spreading pubescence, hirtellous or tomentellous.

Common Name: Tall ironweed.
Habitat: Low woods, floodplain forests, open areas.
Range: Pennsylvania to Michigan, south to Missouri and Mississippi.
Illinois Distribution: Common in the southern three-fourths of Illinois, less common
    in the northern one-fourth.

This is the more common variety of *V. gigantea* in Illinois. The spreading hairs on
the lower surface of the leaves, stems, and peduncles distinguish it from the typical
variety.
    Plants that seem to be somewhat intermediate between the two varieties of *V.
gigantea* in Illinois may be the hybrid known as *V. X illinoensis*.
    Variety *taeniotricha* flowers from July to October.

4. **Vernonia missurica** Raf. Herb. Raf. 28. 1833.
Vernonia altissima Nutt. var. *grandiflora* Gray, Lyn. Fl. 1:90. 1884.

Perennial herbs with rhizomes; stems stout, branched, erect, to 1.2 m tall, densely
tomentose; leaves cauline, alternate, lanceolate to oblong, acuminate at the apex,
tapering to the short-petiolate or sessile base, to 20 cm long, to 5 cm wide, serrate,
scabrous on the upper surface, tomentose or tomentellous on the lower surface; heads
many, in corymbs, discoid, on short peduncles; involucre campanulate or ovoid; phyllaries many, in 6–7 series, appressed or slightly spreading, obtuse to acute to apiculate at

the apex, ciliolate, purplish, the outer lanceolate, 1–2 mm long, the inner oblong, 6–9 mm long; receptacle flat, epaleate; ray flowers absent; disc flowers numerous per head, tubular, bisexual. fertile, the corolla 5-lobed; cypselae cylindric, obscurely ribbed, usually pubescent, 3.5–4.5 mm long; pappus in 2 series, stramineous, the outer of 25–30 scales 0.5–1.0 mm long, the inner of about 40 capillary bristles 6–8 mm long.

Common Name: Missouri ironweed.
Habitat: Low, open woods, prairies.
Range: Michigan to Iowa to Nebraska, south to Texas and Georgia.
Illinois Distribution: Common throughout Illinois.

This is the most common and widespread species of *Vernonia* in Illinois. It differs from the similarly pubescent *V. baldwinii* in its obtuse to apiculate, appressed or slightly spreading phyllaries.

Rose-flowered plants have been called f. *carnea* by Standley. White-flowered plants have been called f. *swinkii* by Steyermark.

Vernonia missurica flowers from July to September.

5. **Vernonia baldwinii** Torr. Ann. Lyc. Nate. Hist. N. Y. 2:211. 1827.

Perennial herbs with rhizomes; stems erect, branched, to 1.2 m tall, densely tomentose; leaves cauline, alternate, lanceolate to oblong-lanceolate, acute to acuminate at the apex, tapering to the short-petiolate base, to 15 cm long, to 4.5 cm wide, glandular-dotted, serrate, tomentose or tomentellous on the lower surface; heads numerous, in corymbs, discoid, on short, tomentose peduncles; involucre campanulate to hemispheric; phyllaries many, in 5–6 series, abruptly acuminate at the apex, ciliolate, the outer lance-ovate, 1–2 mm long, the inner oblong to lanceolate, to 8 mm long; receptacle flat, epaleate; ray flowers absent; disc flowers up to 40 per head, tubular, bisexual, fertile, purple, the corolla 5-lobed; cypselae cylindric, pubescent, 8- or 10-ribbed, 2.5–3.0 mm long; pappus purplish, in 2 series, the outer of 25–30 scales up to 1 mm long, the inner of up to 40 capillary bristles 5–7 mm long.

Two varieties occur in Illinois. Most specimens of *V. baldwinii* may be separated into varieties based on the tips of the phyllaries. However, a few specimens are intermediate and are difficult to place. Both varieties recognized here are scattered, mostly in the southern half of Illinois.

a. Phyllaries recurving at the apex . . . . 5a. *V. baldwinii* var. *baldwinii*
a. Phyllaries more or less appressed at the apex. . . . . .5b. *V. baldwinii* var. *interior*

5a. **Vernonia baldwinii** Raf. var. **baldwinii**
Phyllaries recurving at the apex.

Common Name: Baldwin's ironweed; western ironweed.
Habitat: Prairies, open ground.
Range: Michigan to Nebraska, south to Texas, Louisiana, and Kentucky.
Illinois Distribution: Occasional in the southern two-thirds of Illinois; also DuPage, Kane, and Lake counties.

Variety *baldwinii* flowers from July to September.

5b. **Vernonia baldwinii** Raf. var. **interior** (Small) B. G. Schub. Rhodora 38:370. 1936.

Vernonia interior Small, Bull. Torrey Club 27:279. 1900.

Vernonia interior Small var. *baldwinii* (Raf.) Mack. & Bush, Fl. Jackson Co., Mo. 190. 1903.

Vernonia baldwinii Raf. ssp. *interior* (Small) W. Z. Faust, Brittonia 24:377. 1972.

Phyllaries more or less appressed at the apex.

Common Name: Western ironweed.
Habitat: Prairies, open ground.
Range: Illinois to Nebraska and Colorado, south to Texas, Louisiana, and Kentucky; Michigan.
Illinois Distribution: Scattered in the southern half of Illinois.

This variety is not always distinguishable from var. *baldwinii*.

Variety *interior* flowers from July to September.

Tribe Cichorieae Lam. & DC.

Herbaceous (in Illinois) annuals, biennials, or perennials; latex present; leaves basal and/or cauline, alternate, usually toothed or pinnately lobed, sometimes prickly; heads liguliferous, borne singly or in corymbs or panicles, with 1–15 bractlets, or bractlets absent; involucre campanulate or cylindric; phyllaries several to many, usually in 3–5 series, usually free, unequal; receptacle flat or convex, paleate or epaleate; flowers ligulate, bisexual, fertile, the corolla 5-lobed, of various colors, zygomorphic; cypselae of various shapes, sometimes flattened, smooth or rugose or muricate or tuberculate, often ribbed, glabrous or pubescent, sometimes beaked; pappus of barbellate bristles or plumose bristles, or scales, or a combination of these, rarely absent.

There are about 100 genera and 1,600 species in this tribe, found throughout the world.

This tribe is distinguished by its all ligulate flowering heads and the presence of latex. There are compelling reasons to recognize this tribe as a separate family.

Twenty-one genera occur in Illinois.

1. Flowering heads blue or purple or pinkish.
    2. Cypselae beakless.
        3. Pappus a crown of 2 to many scales in 2 to several series.
            4. Pappus of 12 or more blunt scales in 2–3 series; cauline leaves present . . . .
            . . . . 100. *Cichorium*
            4. Pappus of 2–5 aristate scales in several series; cauline leaves absent . . . .
            . . . . 101. *Catananche*
        3. Pappus of capillary bristles in 1 series . . . . 109. *Nabalus*
    2. Cypselae with a beak about 0.5 mm long or filiform and up to 6 mm long.
        5. Pappus of capillary bristles; heads several; peduncles subtended by bractlets.
            6. Beak of cypselae about 0.5 mm long; pappus in 1 series . . . . 107. *Mulgedium*
            6. Beak of cypselae usually 1–6 mm, long, filiform; pappus usually in 2 or more series . . . . 108. *Lactuca*

5. Pappus of plumose bristles; head solitary; peduncles not subtended by bractlets . . . . 116. *Tragopogon*
1. Flowering heads yellow, orange, cream, or whitish.
  7. Flowering heads cream or whitish . . . . 109. *Nabalus*
  7. Flowering heads yellow or orange.
    8. Cypselae with a beak 0.5–12.0 mm long.
      9. Cypselae with a stout beak about 0.5 mm long.
        10. Pappus of capillary bristles; heads usually numerous; bractlets at base of peduncle up to 10 . . . . 108. *Lactuca*
        10. Pappus of plumose bristles; heads 1 to very few; bractlets at base of peduncle up to 20 . . . . 112. *Leontodon*
      9. Cypselae with a filiform beak 1–12 mm long.
        11. Receptacle paleate; pappus of 2 kinds, the outer of capillary bristles, the inner of plumose bristles . . . . 113. *Hypochaeris*
        11. Receptacle epaleate; pappus not as above.
          12. Outer pappus of scales; inner pappus of capillary bristles; flowering heads 1 to few . . . . 120. *Pyrrhopappus*
          12. Pappus of all bristles; flowering heads one or numerous.
            13. Pappus of capillary bristles.
              14. Stems scapose . . . . 103. *Taraxacum*
              14. Stems with some cauline leaves . . . . 104. *Chondrilla*
            13. Pappus of plumose bristles.
              15. Flowering head solitary, 4–8 cm across. . . . . 116. *Tragopogon*
              15. Flowering heads numerous, up to 2 cm across.
                16. Peduncles subtended by 5 foliaceous bracts . . . . 114. *Helminotheca*
                16. Peduncles subtended by numerous narrow bracts . . . 115. *Picris*
    8. Cypselae beakless.
      17. Leaves with spinescent teeth; pappus in 4 series . . . . 110. *Sonchus*
      17. Leaves not spinescent; pappus in 1–2 series, or absent.
        18. Pappus absent.
          19. Flowering head usually solitary; peduncles not subtended by bractlets . . . . 119. *Serinia*
          19. Flowering heads few to several; peduncles subtended by 4–5 bractlets . . . . 105. *Lapsana*
        18. Pappus present.
          20. Outer pappus of scales, inner pappus of bristles.
            21. Phyllaries up to 35 in 2–5 series; cypselae with 10 ribs . . . . 117. *Nothocalais*
            21. Phyllaries up to 18 in 1–2 series; cypselae with 10–20 ribs . . . 118. *Krigia*
          20. Pappus of all bristles.
            22. Pappus of plumose bristles . . . . 112. *Leontodon*
            22. Pappus of capillary bristles.
              23. None of the leaves pinnatifid . . . . 111. *Hieracium*
              23. Some of the leaves pinnatifid.
                24. Peduncles subtended by 5–12 bractlets; involucre 4–15 mm across; phyllaries up to 16 per head. . . . . 102. *Crepis*
                24. Peduncles subtended by 3–5 bractlets; involucre 2–3 mm across; phyllaries 8 per head. . . . . 106. *Youngia*

Following is a summary of characteristics of the genera of Cichorieae in Illinois:

Flowers blue. *Catananche, Cichorium, Lactuca, Mulgedium.*

Flowers purple or purplish or pinkish. *Nabalus, Tragopogon.*

Flowers white or cream. *Nabalus.*

Flowers yellow or orange. *Lactuca, Nabalus, Sonchus, Serinia, Lapsana, Hieracium, Crepis, Chondrilla, Taraxacum, Pyrrhopappus, Nothocalais, Youngia, Krigia, Hypochaeris, Leontodon, Picris, Helminotheca, Tragopogon.*

Cypselae beakless. *Catananche, Cichorium, Nabalus, Sonchus, Serinia, Lapsana, Leontodon, Hieracium, Crepis, Nothocalais, Youngia, Krigia.*

Cypselae with a beak to 1 mm long. *Mulgedium, Lactuca, Leontodon.*

Cypselae with a beak more than 1 mm long. *Lactuca, Chondrilla, Taraxacum, Pyrrhopappus, Hypochaeris, Picris, Helminotheca, Tragopogon.*

Pappus absent. *Lapsana, Serinia.*

Pappus only a crown of scales. *Catananche, Cichorium.*

Pappus of outer scales and inner plumose bristles. *Leontodon.*

Pappus of outer scales and inner capillary bristles. *Pyrrhopappus, Nothocalais, Krigia.*

Pappus of outer barbellate bristles and inner plumose bristles. *Hypochaeris.*

Pappus only of capillary bristles. *Lactuca* in 2–3 series, *Mulgedium* in 1 series, *Nabalus* in 1 series, *Sonchus* in 4 series, *Hieracium* in 1–2 series, *Crepis* in 2 series, *Chondrilla* in 1 series, *Taraxacum* in 1 series, *Youngia* in 1 series.

Pappus only of plumose bristles. *Leontodon* in 2 series, *Picris* in 2–3 series, *Helminotheca* in 2 series, *Tragopogon* in 1 series.

Phyllaries usually 8 or 9 per head. *Serinia* in 1–2 series, *Lapsana* in 1 series, *Chondrilla* in 1 series, *Youngia* in 2–3 series.

Phyllaries usually 10–12 per head. *Lactuca* in 2 series, *Mulgedium* in 1–2 series, *Nabalus* in 1 series, *Picris* in 1 series, *Helminotheca* in 1 series.

Phyllaries usually more than 12 per head. *Catananche* in several series, *Cichorium* in 2–3 series, *Nabalus* in 1 series, *Sonchus* in 3–5 series, *Hieracium* in 2 series, *Crepis* in 1–2 series, *Taraxacum* in 2 series, *Pyrrhopappus* in 2 series, *Nothocalais* in 2–5 series, *Krigia* in 1–2 series, *Hypochaeris* in 3–4 series, *Leontodon* in 2 series, *Tragopogon* in 1 series.

Receptacle paleate. *Hypochaeris.*

Flowering head always solitary. *Pyrrhopappus, Nothocalais, Krigia, Hypochaeris, Leontodon, Tragopogon, Serinia.*

Bractlets absent at base of peduncles. *Sonchus, Serinia, Nothocalais, Krigia, Hypochaeris, Tragopogon.*

Bractlets usually up to 5 at base of peduncles. *Nabalus, Lapsana, Hieracium, Chondrilla, Youngia, Helminotheca* (foliaceous).

Bractlets usually more than 5 at base of peduncles. *Catananthe, Cichorium, Lactuca, Mulgedium, Nabalus, Hieracium, Crepis, Taraxacum, Pyrrhopappus, Leontodon, Picris.*

It is extremely important to observe the cypselae and pappus when distinguishing the genera in this tribe.

100. **Cichorium** L.—Chicory

Perennial herbs from a taproot; latex present; stems erect, branched; leaves basal and cauline, the basal deeply pinnately lobed or toothed, the cauline alternate, toothed or entire; heads ligulate, borne in glomerules, sessile or on peduncles without bractlets at the base; involucre cylindric; phyllaries up to 15, in 2 series, the outer usually spreading; receptacle flat, epaleate, pitted; ligules up to 30 per head, bisexual, fertile, blue or rarely white or rose; cypselae angular or ribbed, not beaked; pappus of short, white, blunt scales in 2–3 series.

Six species are in the genus, all native to the Old World.

1. **Cichorium intybus** L. Sp. Pl. 2:81. 1753.

Perennial herbs from a taproot; latex present; stems erect, branched, glabrous or sometimes pubescent, to 1.5 m tall; leaves basal and cauline, the basal spatulate, acute at the apex, tapering to the usually petiolate base, shallowly lobed to dentate, to 30 cm long, to 10 cm wide, usually glabrous, the cauline alternate, simple, lanceolate to linear, acute at the apex, sessile or sometimes clasping at the base, denticulate or entire, usually glabrous; heads ligulate, up to 2.5 cm across, bisexual, fertile, sessile or pedunculate, without bractlets; involucre cylindric; phyllaries about 15, in 2 series, the outer lanceolate to lance-ovate, 4–7 mm long, somewhat spreading, the inner linear to linear-lanceolate, up to 12 mm long, erect, usually with glandular hairs; receptacle flat, epaleate; ligules up to 25 per head, bisexual, fertile, blue, rarely white or rose; cypselae 5-angled or 5-ribbed, 2–3 mm long, glabrous; pappus a crown of very short scales.

Common Name: Chicory.
Habitat: Disturbed soil, particularly along roads.
Range: Native to Europe and Asia; adventive in most of the United States.
Illinois Distribution: Common throughout the state; probably in every county.

This is a common plant of old fields and roadsides. The flowering heads are usually closed by noon. In the past, the roots have been roasted and ground and used as a substitute for coffee.

White- and rose-flowered plants have been found in Illinois. They have been named as forms. The white one is f. *album*; the rose one is f. *roseum*.

Cichorium intybus flowers from June to November.

101. **Catananche** L.—Cupid's Dart

Annual or perennial (in ours) herbs; latex present; leaves all basal, numerous, linear to lanceolate to oblanceolate, acute at the apex, tapering to the base, tomentose or glabrous, entire or toothed; cauline leaves absent; heads ligulate, borne on slender peduncles, without bractlets at base; involucre cylindric; phyllaries numerous, in several series, scarious in the upper two-thirds; receptacle epaleate but bristly; ligules flat, toothed at apex, bisexual, fertile, yellow or blue (in ours); cypselae oblongoid, pubescent, the pappus of 2–5 aristate scales.

This genus comprises 5 species, all native to the Mediterranean area. The following ornamental is rarely escaped in the United States.

1. **Catananche caerulea** L. Sp. Pl. 2:812. 1753.

Perennial herbs to 60 cm tall; latex present; leaves all crowded near base of plant, numerous, lanceolate to oblanceolate, 2–3 cm long, 2–5 mm wide, entire or sparsely toothed, tomentose; heads ligulate, up to 4 cm across, on slender peduncles, without bractlets at base; involucre cylindric; phyllaries numerous, in several series, scarious in the upper two-thirds; ligules 11–18 mm long, toothed at apex, blue; cypselae oblongoid, 2–3 mm long, the pappus of 2–5 aristate scales.

Common Name: Cupid's dart.
Habitat: Garden escape.
Range: Native to the Mediterranean area. This is the first report of a spontaneous collection in the United States.
Illinois Distribution: Known only from DeKalb County.

This species differs from all others in the tribe Cichorieae in the 2–5 aristate scales that compose the pappus.

The only collection was made by Paul Sorensen where it apparently persisted as a garden escape.

This species flowers during July.

102. **Crepis** L.—Hawksbeard

Annual, biennial, or perennial herbs with taproots or rhizomes; latex present; stems branched or unbranched, often hispid; leaves basal and cauline, the basal usually in a rosette, often runcinate, pinnately lobed or toothed, on usually winged petioles, the cauline alternate, variously lobed or entire; heads liguliform, borne singly or in cymes or corymbs, subtended by up to 12 unequal bractlets; involucre campanulate or cylindric; phyllaries up to 18 per head, in 1 or 2 series, mostly equal, usually with scarious margins; receptacle flat or convex, usually epaleate, pitted; ligules up to 30 per head, bisexual, fertile, yellow or orange; cypselae variously shaped and colored, curved, ribbed, without a beak, glabrous or pubescent; pappus of numerous barbellate bristles in 1 or 2 series.

Crepis consists of nearly 200 species native to North America, Europe, Asia, and Africa. It differs from *Chondrilla* in its more numerous bractlets at the base of the peduncles, its more numerous ligules per head, and its beakless cypselae.

Three species have been found in Illinois.
1. Cypselae pale brown or tawny; inner phyllaries glabrous.
  2. Involucre 5–8 mm high; cypselae 1.5–2.5 mm long . . . . 1. *C. capillaris*
  2. Involucre 8–12 mm high; cypselae 4–6 mm long . . . . 2. *C. pulchra*
1. Cypselae dark purple-brown; inner phyllaries pubescent on the inner surface . . . .
. . . . 3. *C. tectorum*

1. **Crepis capillaris** (L.) Wallr. Linnaea 14:657. 1840.
Lapsana capillaris L. Sp. Pl. 2:812. 1753.

Annual or perennial herbs with a taproot; latex present; stems ascending to erect, branched, to 60 cm tall, glabrous or hispid; leaves basal and cauline, the basal spatulate to oblanceolate, obtuse to acute and mucronate at the apex, tapering to the petiolate base, to 30 cm long, to 4 cm wide, pinnatifid or coarsely dentate, glabrous or sparsely hispid, the petioles clasping at the base, the cauline alternate, oblong to lanceolate, obtuse to acute at the apex, tapering to the auriculate base, usually entire, glabrous or sparsely hispid; heads liguliform, numerous, in corymbs, up to 2 cm across, subtended by 8 linear, usually glandular and tomentose bractlets 2–4 mm long; involucre cylindric, 5–8 mm high; phyllaries up to 16 per head, in 1 or 2 series, 6–7 mm long, equal, with scarious margins, glabrous on the inner surface, black-setose on the outer surface; receptacle flat or convex, epaleate, pitted; ligules up to 60 per head, bisexual, fertile, yellow, 5-toothed at the apex; cypselae fusiform, pale brown to tawny, 1.5–2.5 mm long, glabrous, 10-ribbed; pappus of numerous white capillary barbellate bristles 3–4 mm long.

Common Name: Smooth hawksbeard.
Habitat: Disturbed soil.
Range: Native to Europe; adventive in much of the United States.
Illinois Distribution: Known from Cook, DuPage, and Lake counties.

This species is distinguished from the other species of *Crepis* in Illinois by its shorter involucres, shorter pale brown or tawny cypselae, and the presence of black setae on the outer face of the phyllaries.
    Crepis capillaris flowers during June and July.

    2. **Crepis pulchra** L. Sp. Pl. 2:806. 1753.
    Annual herbs with a taproot; latex present; stem erect, solitary, branched above, to 1.2 m tall, usually hispid and glandular; leaves basal and cauline, the basal runcinate, obtuse to acute at the apex, tapering to the petiolate base, to 20 cm long, to 4 cm wide, viscid-glandular on both surfaces, the cauline alternate, lanceolate, acute at the apex, tapering to the sessile base, entire, viscid-glandular; heads liguliform, numerous in corymbs, to 4 mm across, subtended by 5–7 glabrous bractlets 1–2 mm long; involucre cylindric, 8–12 mm high; phyllaries up to 14, in 1–2 series, 8–10 mm long, equal, with scarious margins, glabrous on both surfaces; receptacle flat or convex, epaleate, pitted; ligules up to 30 per head, bisexual, fertile, yellow, the corolla 5-toothed at the apex; cypselae more or less cylindric, pale brown or tawny, 4–6 mm long, glabrous, 10-ribbed; pappus of numerous white capillary barbellate bristles 4–5 mm long.

Common Name: Pretty hawksbeard; small-flowered hawksbeard.
Habitat: Disturbed soil.
Range: Native to Europe and Asia; adventive in the United States, mostly in the
    southeastern states.
Illinois Distribution: Known only from Alexander and Johnson counties.

This species is distinguished by its smaller flowering heads, its tall involucre, and its pale brown or tawny cypselae that are 4–6 mm long.
    Crepis pulchra flowers during May and June.

3. **Crepis tectorum** L. Sp. Pl. 2:807 1753.

Annual herbs with a taproot; latex present; stems erect, branched, to 1 m tall, tomentose or hispid; leaves basal and cauline, the basal lanceolate, acute at the apex, tapering to the petiolate base, to 15 cm long, to 4 cm wide, runcinate to dentate, rarely entire, glabrous or sometimes tomentose on the lower surface, the cauline alternate, linear, entire, sessile at the auriculate base; heads liguliform, several to numerous in panicles or corymbs, to 1.8 cm across, subtended by up to 12 subulate bractlets up to 5 mm long; involucre more or less campanulate, 6–10 mm high; phyllaries up to 15, in 1–2 series, equal, 5–10 mm long, with scarious margins, pubescent on both surfaces; receptacle flat or convex, epaleate, pitted; ligules many per head, bisexual, fertile, yellow, the corolla 5-toothed at the apex; cypselae fusiform, dark purple-brown, 3–4 mm long, with 10 scabrous ribs; pappus of numerous white capillary barbellate bristles 4–5 mm long.

Common Name: Narrow-leaved hawksbeard.
Habitat: Disturbed soil, particularly along roadsides.
Range: Native to Europe; adventive in much of the northern half of North America.
Illinois Distribution: Northeastern counties; also Carroll and Macon counties.

This species is readily recognized by its dark purple-brown cypselae with scabrous ribs.

Crepis tectorum tends to be aggressive in the northeastern counties, where it is rapidly spreading.

This species flowers from May to August.

103. **Taraxacum** F. H. Wiggers—Dandelion

Perennial herbs from a taproot; latex present; stems ascending, sometimes scapose; leaves basal, usually in rosettes, runcinate to pinnately lobed; head liguliform, solitary, subtended by up to 20 unequal bractlets; involucre cylindric to campanulate; phyllaries up to 25, in 2 series, more or less equal, spreading during fruiting; receptacle flat, epaleate; ligules numerous per head, bisexual, fertile, yellow, the corolla 5-lobed at the apex; cypselae mostly obovoid, flattened, beaked, several-ribbed, glabrous; pappus of numerous white capillary barbellate bristles in 1 series.

Taraxacum differs from all other Illinois genera in the tribe Cichorieae in its beaked cypselae, its white pappus, and its phyllaries in two series.

There may be 60 species in the genus, found nearly worldwide. Two are known from Illinois.

1. Cypselae olive or olive-brown; terminal lobe of leaves noticeably longer than the other lobes . . . . 1. *T. officinale*
1. Cypselae red to reddish brown; terminal lobe of leaves not noticeably longer than the other lobes . . . . 2. *T. erythrospermum*

1. **Taraxacum officinale** F. H. Wiggers, Prin. Fl. Holsat. 56. 1780.
Leontodon taraxacum L. Sp. Pl. 2:798. 1753.
Leontodon vulgaris Lam. Fl. Fr. 2:113. 1778.
Taraxacum vulgare (Lam.) Schrank, Baier Fl. 11: 1789.
Taraxacum dens-leonis Desf. Fl. Atlant. 2:228. 1800.

Perennial herbs from a taproot; latex present; stems scapose, hollow, up to 10 cm tall, usually glabrous; leaves in a basal rosette, oblong to spatulate, pinnatifid, to 30 cm long, to 10 cm wide, the terminal lobe much longer than the lateral lobes, acute at the apex, tapering to the slightly winged petiole, glabrous or sometimes puberulent on the veins beneath; head liguliform, solitary, to 4 cm across, subtended by up to 18 reflexed bractlets, in 2 series, up to 12 mm long; involucre campanulate; phyllaries up to 18 in 2 series, lanceolate, the margins scarious; receptacle flat, epaleate; ligules numerous per head, bisexual, fertile, yellow, to 20 (−22) mm long, erose at the apex; cypselae fusiform, olive or olive-brown, 2–4 mm long, with up to 12 glabrous or tuberculate ribs, with a slender apical beak up to 10 mm long; pappus of numerous white or sordid capillary barbellate bristles up to 8 mm long.

Common Name: Dandelion.
Habitat: Disturbed soil.
Range: Native to Europe; known from all of the United States except Hawaii.
Illinois Distribution: Very common throughout the state; known from every county.

This pernicious weed is abundant in disturbed soil throughout the state. There is some variation in the degree of lobing of the leaves.

   Taraxacum officinale flowers from February to December.

   2. **Taraxacum erythrospermum** Andrz. ex Besser, Enum. Pl. Vilh. 75. 1822.
Leontodon erythrospermum (Andrz. ex Besser) Eichw. Naturh. Skizze Litth. 150. 1830.
Taraxacum laevigatum (Willd.) DC. var. *erythrospermum* (Andrz. ex Besser) J. Weis,
   Syn. Deut. Schweiz. Fl., ed. 3, 1656. 1900.

Perennial herbs from a taproot; latex present; stems scapose, hollow, to 15 cm tall, usually glabrous; leaves in a basal rosette, oblanceolate to obovate, pinnatifid, acuminate at the apex, tapering to a narrowly winged petiole, the terminal lobe narrower than the lateral lobes, to 15 cm long, to 4 cm wide, usually glabrous except for the veins beneath; head liguliform, solitary, to 4 cm across, subtended by up to 18 reflexed bractlets in 2 series, the bractlets up to 10 mm long; involucre more or less campanulate; phyllaries up to 18 in 2 series, linear-lanceolate, the margins scarious; receptacle flat, epaleate; ligules up to 75 per head, bisexual, fertile, yellow, the outer sometimes with a purple stripe on the back, to 15 mm long, erose at the apex; cypselae oblanceoloid, red or reddish brown, 2.5–3.5 mm long, with about 15 tuberculate ribs, with a slender apical beak up to 8 mm long; pappus of numerous white or sordid capillary barbellate bristles up to 7 mm long.

Common Name: Red-seeded dandelion.
Habitat: Disturbed soil, particularly in sand.
Range: Native to Europe; adventive in most of the United States and Canada.
Illinois Distribution: Occasional throughout Illinois.

This species is readily distinguished by its brick-red cypselae. The terminal lobe of the leaves is usually long and narrow. The leaves are usually more deeply pinnatifid than those of *T. officinale*. Since most people do not collect dandelions, there may be more plants of *T. erythrospermum* in Illinois than the records indicate.

This species flowers from February to December.

104. **Chondrilla** L.—Skeletonweed

Perennial herbs from a taproot; latex present; stems ascending to erect, branched; leaves basal and cauline, the basal pinnatifid to runcinate, petiolate, the cauline alternate, much smaller, entire to denticulate; heads liguliform, 1 to few in a cluster, subtended by 3–4 minute bractlets; involucre cylindric; phyllaries up to 9, in 1 series, equal; receptacle flat, epaleate, pitted; ligules up to 15 per head, bisexual, fertile, yellow, erose at the apex; cypselae oblongoid to linear to cylindric, tan or black, ribbed, with an apical beak; pappus of up to 50 free white capillary persistent bristles.

There are about 25 species in this genus, all native to Europe, Asia, and North Africa.

Only the following adventive species occurs in Illinois.

1. **Chondrilla juncea** L. Sp. Pl. 2:796. 1753.

Perennial herbs from a taproot; latex present; stems ascending to erect, much branched, stiff, to nearly 1 m tall, hirsute near the base; leaves basal and cauline, the basal pinnatifid to runcinate, obtuse to acute at the apex, tapering to the petiolate base, to 12 cm long, to 3 cm wide, glabrous or hidpidulous, usually withered by flowering time, the cauline alternate, much smaller, linear to linear-lanceolate, acute at the apex, tapering to the sessile base, entire or denticulate; heads liguliform, usually several in clusters, up to 1.5 cm across, subtended by 3–4 minute bractlets; involucre cylindric; phyllaries usually 9, in 1 series, equal, tomentose; receptacle flat, epaleate, pitted; ligules up to 15 per head, bisexual, fertile, yellow, erose at the apex; cypselae cylindric, 3–4 mm long, usually tan, ribbed, muricate, at least near the apex, with an apical beak 5–6 mm long; pappus of numerous white capillary bristles in 1 series, 5–6 mm long, persistent.

Common Name: Rush skeletonweed.
Habitat: Disturbed soil.
Range: Native to Europe, Asia, and North Africa; adventive in a few U.S. states.
Illinois Distribution: Jackson and St. Clair counties.

This species is the only member of the Asteraceae in Illinois with yellow flowering heads, beaked cylindric cypselae, presence of cauline leaves, and pappus consisting only of capillary bristles.

Chondrilla juncea flowers from July to September.

105. **Lapsana** L.—Nipplewort

Annual herbs with fibrous roots; latex present; stems erect, branched; leaves basal and cauline, the basal often lobed with a winged petiole, the cauline alternate, smaller, often sessile; heads liguliform, several, in corymbs, subtended by 4–5 very tiny bractlets; involucre cylindric; phyllaries 8, in 1 series, nearly equal, without scarious margins; receptacle flat, epaleate; ligules up to 15 per head, bisexual, fertile, yellow, erose at the apex; cypselae of 2 different lengths, tan or brown, oblongoid, usually not flattened, several-ribbed, without a beak; pappus absent.

Lapsana is now considered to be a genus with only 1 species native to Europe and Asia. Several species in the past that were included in *Lapsana* are now placed in *Lapsanastrum*.

1. **Lapsana communis** L. Sp. Pl. 2:811. 1753.

Annual herbs with fibrous roots; latex present; stems erect, branched, to 1 m tall, glabrous above, hispid below; leaves basal and cauline, the basal ovate, acute at the apex, tapering to the narrowly winged petiole, to 12 cm long, to 5 cm wide, dentate and often with 2–6 lobes, glabrous or sparsely hirsute, at least on the lower surface, the cauline lanceolate to oblong, acute at the apex, tapering to the sessile base, smaller than the basal, usually entire, often glabrous; heads liguli-form, few to several in a corymb, to 1.2 cm across, subtended by 4–5 bractlets up to 1 mm long; involucre cylindric; phyllaries 8, in 1 series, narrowly oblong, nearly equal, keeled, without a scarious margin, to 10 mm long; ligules up to 15 per head, bisexual, fertile, yellow, erose at the apex, 7–10 mm long; cypselae oblongoid, of two lengths, the outer much longer than the inner, tan or brown, usually not flattened, with about 20 ribs, glabrous; pappus absent.

Common Name: Nipplewort.
Habitat: Disturbed soil.
Range: Native to Europe and Asia; introduced into several U.S. states.
Illinois Distribution: Occasional in the northern one-third of Illinois.

This species is distinguished by its yellow flowering heads, its 3–4 tiny bractlets per head, and the absence of pappus.

Lapsana communis flowers from June to September.

106. **Youngia** Cass.—Japanese Hawksbeard

Annual, biennial, or perennial herbs with a taproot; latex present; stems often scapose; leaves mostly basal, pinnately lobed, petiolate; heads liguliform, few to numerous, in corymbs or panicles, subtended by 3–5 usually ovate bractlets; in-volucre cylindric to campanulate, 2–3 mm high; phyllaries 8, in 1–2 series, equal, the margins scarious; receptacle flat or convex, epaleate, pitted; ligules up to 25 per head, bisexual, fertile, yellow; cypselae terete or flattened, without a beak, with 11–13 scabrellous ribs; pappus of 40–60 basally connate capillary bristles in 1 series, persistent.

Youngia consists of about 30 species native to Asia. It differs from *Crepis* in its smaller involucre and fewer bristles on the pappus. The bristles are connate basally.

Only the following adventive species has been found in Illinois.

1. **Youngia japonica** (L.) DC. Prodr. 7:194. 1838.
Prenanthes japonica L. Mant. Pl. 107. 1767.
Crepis japonica (L.) Benth. Fl. Hongk. 194. 1861.

Annual or perennial herbs with a taproot; latex present; leaves mostly basal, usually pinnately lobed, acute at the apex, tapering to the petiolate base, to 15 cm long, to 4 cm wide, glabrous or pubescent; heads liguliform, several in corymbs, to 6 mm across, subtended by 3–5 small ovate bractlets; involucre cylindric to campanulate, 2–3 mm high; phyllaries 8, in 1–2 series, linear to lanceolate, with a scarious margin, glabrous or appressed-pubescent, 3.5–6.0 mm long; receptacle usually flat, epaleate, pitted; ligules up to 20 per head, bisexual, fertile, yellow; cypselae fusiform, usually terete, reddish brown, without a beak, with 11–13 scabrellous ribs, 1.5–2.5 mm long; pappus of 40–60 white capillary barbellate bristles connate at the base, 2.5–3.5 mm long, persistent.

Common Name: Japanese hawksbeard.
Habitat: Disturbed soil.
Range: Native to Asia; adventive from New York to Kentucky, south to Texas and
    Florida; Illinois.
Illinois Distribution: Edge of a parking lot in Metropolis, Massac County.

This weed is very common in the southeastern United States. It is recognized by its basally connate pappus bristles, its smaller involucre, and its beakless cypselae. The only Illinois collection was made by the author and has been deposited in the herbarium of the Missouri Botanical Garden.

Youngia japonica flowers during June and July.

107. **Mulgedium** Cass. in F. Cuv.—Blue Lettuce

Perennial herbs with rhizomes; latex present; stems erect, branched, usually glabrous; leaves basal and cauline, the basal petiolate or sessile, entire to dentate to pinnately lobed, glabrous, the cauline alternate, entire, sessile; heads liguliform, in corymbs or panicles, subtended by up to 12 bractlets; involucre cylindric; phyllaries up to 12 (–13), in 1–2 series, usually equal, without a scarious margin; receptacle flat, epaleate, pitted; ligules up to 35 (–50) per head, bisexual, fertile, blue; cypselae lanceolate, flat, reddish brown, several-ribbed, glabrous, without a beak; pappus of numerous white capillary barbellate bristles or nearly smooth bristles, persistent.

Mulgedium is a genus of 15 species native to Europe and Asia and 1 in North America.

Species of *Mulgedium* have often been placed in *Lactuca*. They differ in being perennial herbs with rhizomes, lacking a beak on the cypselae, and having nonscarious phyllaries. *Lactuca* species have beaked cypselae and are annuals or biennials. However, *L. biennis* and *L. floridana* seem to be intermediate between *Mulgedium* and *Lactuca*. These two species have very short, stout beaks on the cypselae, while the other species of *Lactuca* have filiform beaks. It may be best to treat these 2 species in the genus *Cicerbita* Wallr., although I have not done so in this work.

Only the following species occurs in Illinois.

1. **Mulgedium pulchellum** (Pursh) G. Don in R. Sweet, Hort. Brit., ed. 3, 418. 1839.
Sonchus pulchellus Pursh, Fl. Am. Sept. 2:502. 1813.
Lactuca pulchella (Pursh) DC. Prodr. 7:134. 1838.
Lactuca tatarica L. ssp. *pulchella* (Pursh) Stebbins, Madrono 5:123. 1939.
Lactuca tatarica L. var. *pulchella* (Pursh) Breitung, Can. Field Nat. 71:70. 1957.

Perennial herbs with rhizomes; latex present; stems erect, branched, to 1 m tall, glabrous, sometimes glaucous; leaves basal and cauline, the basal narrowly lanceolate to oblong, acuminate at the apex, tapering to the usually petiolate base, to 15 cm long, to 3 cm wide, entire, dentate, or often pinnately lobed, glabrous, sometimes glaucous, the cauline linear-lanceolate, acute at the apex, tapering to the sessile base, entire, glabrous; heads liguliform, numerous, up to 3 cm across, subtended by up to 12 lanceolate bractlets; involucre campanulate; phyllaries up to 12 (–13), in 1–2 series, nearly equal, without a scarious margin; receptacle flat, epaleate, pitted; ligules up to 35 per head, bisexual, fertile, blue, erose at the apex; cypselae lanceolate, flattened, reddish brown, 4–5 mm long, glabrous, several-ribbed, without a beak or with a beak less than 1 mm long; pappus of numerous white capillary barbellate bristles up to 10 mm long, persistent.

Common Name: Showy blue lettuce.
Habitat: Along railroads and other disturbed areas.
Range: Native to western North America; adventive in Illinois.
Illinois Distribution: Known from the northeastern counties of the state.

This species differs from the blue-flowered species of *Lactuca* in its perennial, rhizomatous habit and the usual absence of a beak on the cypselae.
  Mulgedium pulchellum flowers during July and August.

108. **Lactuca** L.—Lettuce

Annual or biennial herbs with taproots; latex present; stems erect, branched, glabrous or pubescent; leaves basal and cauline, entire to toothed to pinnately lobed, glabrous or pubescent, petiolate or sessile; heads liguliform, borne singly or in corymbs or panicles, subtended by up to 10 bractlets; involucre campanulate or cylindric; phyllaries up to 12, in 2 series, equal, usually with scarious margins; receptacle flat or convex, epaleate, pitted; ligules up to 50 per head, bisexual, fertile, yellow or blue or reddish or salmon-colored; cypselae tan to reddish brown to black to yellowish, more or less flattened, with a terminal beak, ribbed, usually glabrous; pappus of numerous white, brown, cream, or gray capillary bristles, sometimes surrounded by a minute crown of scales.
  Lactuca consists of about 75 species found worldwide. The beaked cypselae and the annual or biennial habit with taproots distinguish this genus from *Mulgedium*.

1. Leaves broadly ovate . . . . .9. *L. sativa*
1. Leaves elliptic to linear-lanceolate, often deeply pinnatifid, usually not ovate.
  2. Leaves prickly on the midvein beneath.
    3. Involucre 15–22 mm high; cypselae black, 1- to 3-nerved on each face . . . .
    . . . . 6. *L. ludoviciana*
    3. Involucre 8–15 mm high; cypselae yellow-gray, several-nerved on each face . . . . .
    . . . . 8. *L. serriola*
  2. Leaves not prickly on the midvein beneath or, if with a few prickles, the leaves never
    more than 1 cm wide.
    4. Pappus brown or gray . . . . .1. *L. biennis*
    4. Pappus white or cream.
      5. Involucre 15–22 mm high.
        6. Flowers reddish or salmon-colored; leaves without a prickly margin . . . .
        . . . . .5. *L. hirsuta*
        6. Flowers yellow or blue; leaves often with a prickly margin. . . 6. *L. ludoviciana*
      5. Involucre up to 15 mm high (rarely to 18 mm in *L. saligna*, which has all leaves
        less than 1 cm wide).
        7. None of the leaves more than 1 cm wide; beak of cypselae twice as long as
          the body . . . . .7. *L. saligna*
        7. Some or all of the leaves more than 1 cm wide; beak of cypselae nearly
          absent to almost as long as the body.
          8. Flowers yellow; beak of cypselae at least half as long as to equaling the
            body. . . . . .4. *L. canadensis*
          8. Flowers blue; beak of cypselae nearly absent or less than half as long as
            the body.
            9. Some of the cypselae beakless or nearly so; pappus white; terminal
              lobe of most leaves broadly triangular. . . . . 2. *L. floridana*
            9. Cypselae with a beak one-fourth to half as long as the body; pappus
              cream; terminal lobes of leaves not broadly triangular. . . 3. *L. X morsii*

1. **Lactuca biennis** (Moench) Fern. Rhodora 42:300. 1940.
Sonchus biennis Moench, Meth. 545. 1794.
Sonchus leucophyllus Willd. Sp. Pl. 3:1520. 1804.
Mulgedium leucophaeum (Willd.) DC. Prodr. 7:250. 1838.
Lactuca leucophaea (Willd.) Gray, Proc. Am. Acad. 73. 1883.
Lactuca spicata Hitchc. in Britt. & Rose, Ill. Fl. 3:276. 1898, *non* Lam. (1789).
Lactuca spicata Hitchc. in Britt. & Rose var. *aurea* Jenn. Ann. Carnegie Mus. 13:440.
    1922.
Lactuca biennis (Moench) Fern. f. *aurea* (Jenn.) Fern. Rhodora 42:302. 1940.

    Annual or biennial herbs with a taproot; latex present; stems erect, branched, to
3 m tall, glabrous or nearly so; leaves mostly cauline, alternate, the lowermost runcinate, dentate or less commonly entire, acute at the apex, tapering to the winged
petiole, to 30 cm long, to 15 cm wide, glabrous or sometimes pubescent on the veins
on the lower surface, the middle and upper leaves lanceolate to ovate, pinnatifid to
dentate to entire, acute at the apex, tapering to the sessile and sometimes auriculate
base, glabrous or with setae on the veins on the lower surface; heads liguliform,

numerous, borne in panicles, up to 5 mm across, subtended by up to 10 deltate bractlets; involucre campanulate; phyllaries up to 12, in 2 series, with a narrow scarious margin, reflexed during fruiting; receptacle flat, epaleate, pitted; ligules numerous per head, bisexual, fertile, blue or rarely yellowish; cypselae lanceolate to elliptic to oblong, flattened, brown, 4–5 mm long, with a stout beak up to 5 mm long; pappus of brownish or rarely grayish capillary bristles 4–6 mm long.

Two forms occur in Illinois.

a. Some or all of the leaves lobed . . . . 1a. *L. biennis* f. *biennis*
a. None of the leaves lobed . . . . 1b. *L. biennis* f. *integrifolia*

1a. **Lactuca biennis** (Moench) Fern. f. **biennis**
Some or all of the leaves lobed.

Common Name: Tall blue lettuce.
Habitat: Open woods, floodplain woods, disturbed woods.
Range: Newfoundland to Yukon, south to California, New Mexico, Tennessee, and
    North Carolina.
Illinois Distribution: Occasional throughout the state.

Most specimens of *L. biennis* have some or all of the leaves lobed. Most of the flowering heads are blue, although a few plants with yellowish heads have been seen in Illinois. Plants up to 3 m tall are common. Plants with yellow ligules have been called f. *aurea*.

The very short beak of the cypselae led deCandolle to place *L. biennis* in the genus *Mulgedium*.

This form flowers during August and September.

1b. **Lactuca biennis** (Moench) Fern. f. **integrifolia** (Torr. & Gray) Fern. Rhodora 42:300. 1940.
Mulgedium leucophaeum DC. var. *integrifolium* Torr. & Gray, Fl. N. Am. 2:499. 1893.
Lactuca spicata (Lam.) Hitchc. var. *integrifolia* (Torr. & Gray) Britt. Mem. Torrey Club 5:350. 1894.

None of the leaves lobed, although some of them are sparingly serrulate.

This variant occurs with some regularity throughout the state, where it occupies the same habitats as f. *biennis*. It flowers during August and September.

2. **Lactuca floridana** (L.) Gaertn. Fruct. Sem. Pl. 2:362. 1791.
Sonchus floridanus L. Sp. Pl. 2:974. 1753.
Mulgedium floridanum (L.) DC. Prodr. 7:349. 1791.
Lactuca floridana (L.) Gaertn. f. *leucantha* Fern. Rhodora 42:498. 1940.

Annual or biennial herbs with a taproot; latex present; stems erect, branched, to 3 m tall, glabrous or sometimes villous; leaves alternate, the lower ones usually deeply pinnatifid, acute at the apex, tapering to a winged petiole, to 30 cm long, to 15 cm wide, the terminal lobe broadly triangular with entire or rarely slightly toothed, margins, glabrous or the veins of the lower surface pilose, the middle and upper leaves smaller, usually entire or denticulate, glabrous, sessile; heads liguliform, numerous, in panicles, up to 5 mm across, subtended by up to 10 deltate bractlets; involucre campanulate; phyllaries up to 12, in 2 series, with a scarious margin, reflexed in fruit; receptacle flat, epaleate, pitted; ligules numerous per head, bisexual, fertile, blue, rarely yellow or whitish, erose at the apex; cypselae oblong, flattened, brown, 4–5 mm long, with a stout beak up to 1 mm long; pappus of white capillary bristles 4–5 mm long, persistent.

Two varieties occur in Illinois.
a. All except sometimes the uppermost leaves lobed. . . . . 2a. *L. floridana* var. *floridana*
a. All leaves unlobed, entire or denticulate. . . . . .2b. *L. floridana* var. *villosa*

2a. **Lactuca floridana** (L.) Gaertn. var. **floridana**
Sonchus acuminatus Willd. Sp. Pl. 3:1521. 1804.
Mulgedium acuminatum (Willd.) DC. Prodr. 7:249. 1838.

All except sometimes the uppermost leaves lobed.

Common Name: Woodland lettuce.
Habitat: Woods, savannas, fens, marshes.
Range: New York to Manitoba, south to Texas and Florida.
Illinois Distribution: Common throughout the state.

The most conspicuous characteristic to tell this plant from any other *Lactuca* is the broad triangular terminal lobe of some of the leaves. The white pappus also distinguishes *L. floridana* from *L. biennis*. White-flowered plants may be called f. *leucantha*.

This variety is the more common of the varieties in Illinois. It flowers from July to September.

2b. **Lactuca floridana** (L.) Gaertn. var. **villosa** (Jacq.) Cronq. Rhodora 50:31. 1948.
Lactuca villosa Jacq. Pl. Hort. Schoenbr. 3:62. 1798.
Mulgedium villosum (Jacq.) Small, Fl. Lancaster Co. 316. 1913.

All leaves unlobed, entire or denticulate.
This variety is distinguished from the typical variety by the absence of lobed leaves. Most specimens also have villous stems.

This variety occurs occasionally throughout the state in woods. It flowers from July to September.

3. **Lactuca X morsii** B. L. Robins. Rhodora 1:12–13. 1899.
Annual or biennial herbs from a taproot; latex present; stems erect, branched, to 3 m tall, hirsute near the base; leaves cauline, alternate, some of them runcinate, acute at the apex, tapering to the winged and cordate-clasping base, to 20 cm long, to 10 cm wide, the terminal lobe sometimes more or less triangular with dentate margins, glabrous or sometimes hirsute on the veins below, upper leaves

acuminate at the apex; heads liguliform, numerous, borne in panicles, subtended by about 10 deltate bractlets; involucre campanulate; phyllaries up to 12, in 2 series, 8–10 mm long, purple-tinged, the outer ovate, acute at the apex, the inner oblong, obtuse at the apex; ligules about 18, bisexual, fertile, blue; cypselae lanceolate, flattened, black mottled with brown, 3-ribbed on each face, 4–5 mm long, the filiform beak 1.0–1.5 mm long; pappus of numerous capillary bristles white at first but becoming cream-colored, 4–5 mm long, persistent.

Common Name: Hybrid lettuce.
Habitat: Woods.
Range: Scattered in the eastern United States.
Illinois Distribution: Known only from St. Clair County.

This is reputed to be the hybrid between *L. biennis* and *L. canadensis*, but the more or less triangular terminal lobe of some of the leaves resembles *L. floridana*. The terminal lobe in *L. X morsii* is coarsely dentate, whereas the terminal lobe of the leaves in *L. floridana* is usually entire.

This hybrid flowers during July and August.

4. **Lactuca canadensis** L. Sp. Pl. 2:796. 1753.
Lactuca elongata Muhl. ex Willd. Sp. Pl. 3:1525. 1804.

Annual or biennial herb with a taproot; latex present; stems erect, branched, to 3.5 m tall, glabrous or sometimes hirtellous at the base, occasionally glaucous; leaves cauline, alternate, oblong to lanceolate, sometimes linear, sometimes obovate, acuminate at the apex, tapering to a winged petiole, or sessile and sagittate-auriculate, to 15 cm long, to 8.5 cm wide, entire or denticulate, rarely pinnatifid, glabrous or sometimes pilose on the veins on the lower surface; heads liguliform, numerous, borne in panicles or corymbs, to 10 mm across, subtended by up to 10 deltate bractlets; involucre cylindric; phyllaries 12, in 2 series, the outer shorter than the inner, reflexed in fruit; ligules up to 25 per head, bisexual, fertile, usually yellow, erose at the apex; cypselae oblong to elliptic, more or less flattened, black or dark brown, rugulose, glabrous, 1- or 3-nerved on each face, 5–6 mm long, with a terminal filiform beak up to 3–6 mm long, half as long as to equaling the body; pappus of numerous white capillary bristles 5–7 mm long, persistent.

Lactuca canadensis is distinguished by its usual lack of pinnatifid leaves, its yellow ligules, and the beak of the cypselae being half as long as to equaling the body.

Common Name: Wild lettuce.
Habitat: Dry woods, pastures, prairies, savannas, old fields.
Range: Nova Scotia to Yukon, south to California, Texas, and Florida.
Illinois Distribution: Occasional throughout Illinois.

Four varieties, based on leaf characteristics, have been found in Illinois.
a. All but the lowermost leaves unlobed, sometimes sagittate-auriculate at the base.

b. Cauline leaves entire or nearly so .... 4a. *L. canadensis* var. *canadensis*
b. Cauline leaves regularly denticulate................. 4b. *L. canadensis* var. *obovata*
a. All but sometimes the uppermost leaves pinnatifid, usually not sagittate-auriculate at the base.
    c. Lobes of leaves linear, falcate .... 4c. *L. canadensis* var. *longifolia*
    c. Lobes of leaves broadly obovate, straight or falcate .... 4d. *L. canadensis* var. *latifolia*

4a. **Lactuca canadensis** L. var. **canadensis**
Lactuca sagittifolia Ell. Bot. S. C. & Ga. 2:253. 1821.
Lactuca integrifolia Bigel, Fl. Bost., ed. 2, 287. 1824.

4b. **Lactuca canadensis** L. var. **obovata** Wieg. Rhodora 22:11. 1920.

4c. **Lactuca canadensis** L. var. **longifolia** (Michx.) Farw. Pap. Mich. Acad. Sci. 2:45. 1923.
Lactuca longifolia Michx. Fl. Bor. Am. 2:85. 1803.

4d. **Lactuca canadensis** L. var. **latifolia** Ktze. Rev.Gen. Pl. 1:349. 1891.
Lactuca canadensis flowers from June to September.

5. **Lactuca hirsuta** Muhl. ex Nutt. var. **sanguinea** (Bigelow) Fern. Rhodora 40:481. 1938.
Lactuca sanguinea Bigelow, Fl. Bost., ed. 2, 287. 1824.
Lactuca canadensis L. var. *sanguinea* (Bigelow) Torr. & Gray ex Patterson. Cat. Pl. Ill. 25. 1876.

Biennial herbs from a taproot; latex present; stems erect, branched, to 1.2 m tall, purplish or reddish, nearly glabrous; leaves cauline, alternate, the lowermost usually deeply pinnatifid, acute at the apex, tapering to the winged petiole, to 20 cm long, to 8 cm wide, pilose on both surfaces, the middle and upper usually ovate, unlobed, denticulate, pilose on both surfaces, sessile; heads liguliform, borne in corymbs or panicles, to 6 mm across, subtended by about 10 deltate bractlets; involucre urceolate, 15–22 mm high, usually purplish or reddish; phyllaries up to 12, in 2 series, the outer shorter than the inner, reflexed in fruit; receptacle flat, epaleate, pitted; ligules up to 25 per head, bisexual, fertile, reddish or salmon-colored; cypselae elliptic, more or less flattened, brown, 4–5 mm long, with a filiform beak 2.5–4.0 mm long; pappus of numerous white capillary bristles up to 10 mm long, persistent.

Common Name: Red-stemmed lettuce.
Habitat: Dry woods.
Range: Prince Edward Island to Ontario, south to Texas and Virginia.
Illinois Distribution: Rare in the southern one-sixth of Illinois.

Typical var. *hirsuta*, not yet found in Illinois, has a very hirsute stem. The stems of var. *sanguinea* are purplish or reddish.
    This plant flowers from July to September.

6. **Lactuca ludoviciana** (Nutt.) Riddell, W. J. Med. Phys. Sci. 8:491. 1835.
Sonchus ludovicianus Nutt. Gen. 2:15. 1818.

Lactuca campestris Greene, Pittonia 4:37. 1899.
Lactuca campestris Greene f. *campestris* (Greene) Fern. Rhodora 40:481. 1938.

Biennial herbs from a taproot; latex present; stems erect, branched, to 1.5 m tall, glabrous; leaves cauline, alternate, oblong to ovate, acute at the apex, tapering to the often auriculate-clasping base, to 15 cm long, to 7 cm wide, often pinnatifid with spinulose teeth, glabrous except usually setose on the midvein on the lower surface, glaucous; heads liguliform, several in corymbs, to 10 mm across, subtended by about 10 deltate bractlets; involucre usually cylindric, 15–20 mm high; phyllaries about 15, in 2 series, the outer ovate, the inner narrower, reflexed in fruit; receptacle flat, epaleate, pitted; ligules numerous, bisexual, fertile, yellow or less commonly blue, erose at the apex; cypselae elliptic, more or less flattened, brown or black, 1- or 3-nerved on each face, 4–5 mm long, with a filiform beak 2.5–3.5 mm long; pappus of numerous white capillary bristles up to 10 mm long, usually persistent.

Common Name: Western wild lettuce.
Habitat: Dry prairies.
Range: Wisconsin to British Columbia, south to California, Texas, Louisiana, and
   Kentucky.
Illinois Distribution: Rare in the northern half of Illinois, extending south to Monroe
   County.

The distinguishing features of this species are its involucre 15–20 mm high and its usually spine-tipped teeth of the leaves. Most specimens from Illinois have yellow flowers, but a few blue-flowered plants have been seen. These latter may be called f. *campestris*.
   Lactuca ludoviciana flowers from July to September.

7. **Lactuca saligna** L. Sp. Pl. 2:796. 1753.
   Annual or biennial herbs with a taproot; latex present; stems ascending to erect, branched, to 75 cm tall, glabrous or nearly so; leaves cauline, alternate, the lowermost usually deeply runcinate, acuminate at the apex, sagittate-auriculate at the base, to 15 cm long, to 7 cm wide, usually with spinulose teeth and with setae on the midvein of the lower surface, the middle and upper often unlobed, up to 1 cm wide; heads liguliform, few in a narrow panicle, to 4 mm across, subtended by about 10 bractlets; involucre cylindric, up to 15 (–18) mm high; phyllaries about 15, in 2 series, the outer much shorter than the inner, erect in fruit; receptacle flat, epaleate, pitted; ligules up to 20 per head, bisexual, fertile, yellow, erose at the apex; cypselae oblong to elliptic, more or less flattened, pale brown, 5- or 7-ribbed on each face, 2.5–3.5 mm long, with an apical filiform beak 5–6 mm long; pappus of numerous white capillary bristles 5–6 mm long.

Common Name: Willow-leaved lettuce.
Habitat: Disturbed areas.
Range: Native to Europe; adventive in many of the United States.
Illinois Distribution: Occasional to common throughout the state.

The extremely narrow, prickly-toothed leaves are characteristic for this species.

Lactuca saligna flowers from July to October.

8. **Lactuca serriola** L. Cent. Pl. 2:29. 1756.
Lactuca scariola L. Sp. Pl., ed. 2:1119. 1763.
Lactuca virosa L. var. *integrifolia* S. F. Gray, Nat. Arr. Brit. Pl. 2:417. 1821.
Lactuca scariola L. var. *integrifolia* Bogenh. Fl. Jena 269. 1850.
Lactuca scariola L. var. *integrata* Gren. & Godr. Fl. Fr. 2:320. 1850.
Lactuca serriola L. f. *integrifolia* (S. F. Gray) S. D. Prince & R. N. Carter, Watsonia
    3:37. 1977.

Annual or biennial herbs with a taproot; latex present; stems ascending to
erect, to 2.5 m tall, often prickly, sometimes glaucous; leaves cauline, alternate,
oblong-lanceolate to pinnatifid, rarely entire or merely toothed, acute at the apex,
tapering to the sessile or auriculate-clasping base, to 25 cm long, to 7.5 cm wide,
the margins and the veins beneath prickly, or rarely the prickles absent or sparse;
heads liguliform, numerous, borne in panicles, to 6 mm across, subtended by
about 10 deltate bractlets; involucre cylindric, up to 10 mm high; phyllaries about
15, in 2 series, the outer much shorter than the inner, reflexed in fruit; receptacle
flat, epaleate, pitted; ligules up to 20 per head, bisexual, fertile, yellow, erose at the
apex; cypselae oblanceolate, more or less flattened, gray or tan, 3- to 9-ribbed per
face, 2.5–3.5 mm long, with a filiform beak 2.5–4.0 mm long; pappus of numerous
white capillary bristles 3.5–5.0 mm long, persistent.

Common Name: Prickly lettuce.
Habitat: Disturbed areas.
Range: Throughout most of North America.
Illinois Distribution: Very common throughout the state; in every county.

This common plant of disturbed soil is readily recognized by its prickly leaves,
prickly stems, and yellow flowering heads. Botanists in the past have called this
species *L. scariola*. Plants with unlobed leaves and few or no prickles on the leaves
have been called f. *integrifolia*.

Lactuca serriola flowers from July to November.

9. **Lactuca sativa** L. Sp. Pl. 2:795. 1753.
Annual or biennial herbs from a taproot; latex present; stems ascending to
erect, branched, to 50 (–60) cm tall, glabrous; leaves cauline, alternate, broadly
ovate to orbicular, acute at the apex, tapering to the sessile base, to 20 cm long,
to 10 cm wide, entire or denticulate, prickly toothed, glabrous; heads liguliform,
few to several borne in corymbs; involucre cylindric, up to 12 mm high; phyllaries
about 15, in 2 series, the outer shorter than the inner, erect in fruit; receptacle flat,
epaleate, pitted; ligules up to 25 (–30) per head, bisexual, fertile, yellow, erose at
the apex; cypselae obovate, more or less flattened, gray or tan, with 5–9 nerves per
face, 3–4 mm long, with a filiform beak 3–5 mm long; pappus of numerous white
capillary bristles 3.5–4.5 mm long.

Common Name: Garden lettuce.
Habitat: Disturbed soil.
Range: Native to Europe and Asia; introduced into several U.S. states.
Illinois Distribution: Rarely escaped from gardens and seldom persistent in Illinois.

Lactuca sativa flowers from July to September.

109. **Nabalus** Cass.—White Lettuce

Perennial herbs with a taproot and sometimes with rhizomes; latex present; stems ascending to erect, branched or unbranched; leaves usually basal and cauline, petiolate or less commonly sessile, sometimes auriculate and clasping, entire to dentate to lobed; heads liguliform, borne in racemes, panicles, or corymbs, subtended by 2–12 unequal bractlets; involucre cylindric; phyllaries up to 14, in 1 series, equal, scarious on the margins, glabrous or pubescent; receptacle flat, epaleate; ligules up to 35 per head, bisexual, fertile, white, pink, purplish, or yellowish, erose at the apex; cypselae oblong to fusiform, sometimes angular, 5- to 12-ribbed, usually glabrous, without a beak; pappus of up to 50 white, yellow, tan, or reddish brown capillary bristles, in 1 series.

    Nabalus consists of about 25 North American species. Our species have usually been placed in *Prenanthes*, but I believe *Prenanthes* to be a different genus not found in Illinois. *Prenanthes* occurs in Asia and Africa.

    Five species of *Nabalus* occur in Illinois.

1. Phyllaries glabrous.
  2. Phyllaries 7–10 per head; flowers 8–15 per head; pappus reddish brown. . . 1. *N. albus*
  2. Phyllaries 4–6 per head; flowers 5–6 per head; pappus straw-colored or
    cinnamon-brown. . . . . 2. *N. altissimus*
1. Phyllaries pubescent.
  3. Leaves petiolate; inflorescence corymbose-paniculate. . . . . 4. *N. crepidineus*
  3. Leaves sessile or nearly so; inflorescence narrow and elongate.
    4. Stems and leaves glabrous and usually glaucous; flowers purplish. . . 5. *N. racemosus*
    4. Stems and leaves pubescent, never glaucous; flowers cream-colored . . . 3. *N. asper*

    1. **Nabalus albus** (L.) Hook. Fl. Bor. Am. 1:294. 1833.
Prenanthes alba L. Sp. Pl. 2:798. 1753.

    Perennial herbs with a short taproot; latex present; stems erect, usually branched, up to 1.5 m tall, mostly glabrous, usually glaucous, often purplish; cauline leaves alternate, ovate to deltate, acute at the apex, cordate or truncate at the petiolate base, extremely variable, to 25 cm long, to 15 cm wide, 3-lobed or palmately lobed, denticulate, glabrous on the upper surface, whitened and usually hirsutulous on the lower surface; heads liguliform, several in clusters, corymbose-paniculate, to 3 mm across, subtended by 5–7 lanceolate, glabrous bractlets up to 3 mm long; involucre cylindric to campanulate, 3–5 mm high; phyllaries 7–10, in 1 series, lanceolate, purplish, the margins scarious and ciliate, glabrous, 10–13 mm long; receptacle flat, epaleate; ligules 8–15 per head, bisexual, fertile, white or pinkish; cypselae narrowly oblong, brown or tan, 3.5–6.0 mm long, several-ribbed, without a beak; pappus of up to 50 reddish brown capillary bristles 6–7 mm long.

Common Name: White lettuce.
Habitat: Woods, wooded dune slopes.
Range: Maine to Manitoba, south to South Dakota, Arkansas, and Georgia.
Illinois Distribution: Occasional in the northern two-thirds of Illinois; also Pulaski
    and Union counties.

This species is similar in appearance to *N. altissimus* but has more phyllaries per
head and more flowers per head. The stems of *N. albus* are usually glaucous. It is
similar to *N. altissimus* var. *cinnamomeus* in its reddish brown pappus. The leaves of
N. albus are extremely variable.

  Nabalus albus flowers from August to November.

  2. **Nabalus altissimus** (L.) Hook. Fl. Bor. Am. 1:294. 1833.
Prenanthes altissima L. Sp. Pl. 2:797. 1753.

  Perennial herbs with a thick rhizome; latex present; stems erect, usually
branched, to 2.3 m tall, glabrous or hispidulous, not glaucous, occasionally
purplish; cauline leaves alternate, ovate to deltate, acute at the apex, cordate or
truncate at the petiolate base, extremely variable, to 20 cm long, to 12 cm wide,
entire, dentate, or 3- or 5-lobed, glabrous on the upper surface, usually pubescent
on the veins of the lower surface; heads liguliform, numerous in a panicle or in
axillary clusters, to 4 mm across, subtended by 4–6 black, deltate bractlets 1–4
mm long; involucre cylindric, 2–3 mm high; phyllaries 4–6, in 1 series, linear to
linear-lanceolate, glabrous or pubescent, 10–12 mm long; receptacle flat, epaleate;
ligules 5–6 per head, bisexual, fertile, yellowish; cypselae cylindric, brown or tan,
ribbed, 4–5 mm long, without a beak; pappus of numerous creamy white or reddish brown or cinnamon-colored capillary bristles 5–6 mm long, persistent.

  Two varieties occur in Illinois.
a. Pappus creamy white. . . . . 2a. *N. altissimus* var. *altissimus*
a. Pappus reddish brown or cinnamon-colored. . . . . . . 2b. *N. altissimus* var. *cinnamomeus*

  2a. **Nabalus altissimus** (L.) Hook. var. **altissimus**
Pappus creamy white.

Common Name: Tall white lettuce.
Habitat: Woods.
Range: New Brunswick to Ontario, south to Texas and Georgia.
Illinois Distribution: Occasional in eastern and southern Illinois; also Cook, Lake,
    and Will counties.

The only difference I can detect between this variety and var. *cinnamomeus* is the
creamy white color of the pappus. Both varieties occupy similar habitats and share
similar ranges in Illinois.

  This variety flowers from August to October.

  2b. **Nabalus altissimus** (L.) Hook. var. **cinnamomeus** (Fern.) Mohlenbr. Guide
Ill. Fl., 4th ed., 137. 2013.

Prenanthes altissima L. var. *cinnamomea* Fern. Rhodora 10:95. 1908.

Pappus reddish brown or cinnamon-colored.

The color of the pappus is similar to that of *N. albus*, although it might have a slightly lighter shade. *Nabalus altissimus* var. *cinnamomeus* may be distinguished from *N. albus* by its fewer flowers per head and its fewer phyllaries.

Variety *cinnamomeus* flowers from August to October.

3. **Nabalus asper** (Michx.) Torr. & Gray, Fl. N. Am. 2:483. 1843.
Prenanthes asper Michx. Fl. Bor. Am. 2:83. 1803.
Prenanthes illinoensis Pers. Syn. Pl. 2:366. 1807.

Perennial herbs with thickened taproots; latex present; stems erect, branched or unbranched, to 1.5 m tall, usually purple-speckled, glabrous to hispid and scabrous; leaves basal and cauline, the basal oblong to oblanceolate, obtuse at the apex, tapering to a winged petiole, the cauline alternate, oblong to oblanceolate, acute or obtuse at the apex, tapering to the sessile and sometimes clasping base, to 7 cm long, to 2.5 cm wide, entire or denticulate, not lobed, usually scabrous on the upper surface; heads liguliform, numerous, in narrow racemes, erect to nodding, subtended by up to 12 lanceolate, setose bractlets; involucre cylindric to campanulate, 4–5 mm high; phyllaries up to 10, in 1 series, linear to linear-lanceolate, tan or yellow-green, setose, to 15 mm long; receptacle flat, epaleate; ligules up to 18 per head, bisexual, fertile, cream-colored, up to 15 (–17) mm long; cypselae cylindrical, tan, 10- or 12-ribbed, 5–6 mm long, without a beak; pappus of numerous straw-colored capillary bristles 7–8 mm long, persistent.

Common Name: Rough white lettuce.
Habitat: Dry prairies.
Range: Pennsylvania to South Dakota, south to Oklahoma, Louisiana, and Georgia.
Illinois Distribution: Occasional in the northern two-thirds of Illinois, less common southward.

This is the only species of *Nabulus* in Illinois that has all leaves unlobed. Its scabrous stems and leaves are also distinctive.

Nabulus asper flowers from August to October.

4. **Nabalus crepidineus** (Michx.) DC. Prodr. 7:242. 1838.
Prenanthes crepidinea Michx. Fl. Bor. Am. 2:84. 1803.

Perennial herbs with thick taproots; latex present; stems erect, branched, to 2.5 m tall, glabrous or sometimes puberulent above; leaves basal and cauline, the basal and lower cauline leaves withering early, ovate to deltate, acute at the apex, hastate at the base, with winged petioles, to 20 cm long, to 10 cm wide, entire to dentate to lobed, glabrous or hirsute, the middle and upper cauline leaves similar, on short petioles; heads liguliform, numerous, corymbose-paniculate, nodding, 8–11 mm across, subtended by up to 20 lanceolate or deltate setose bractlets 2–5 mm long; involucre campanulate, up to 14 mm high; phyllaries 10–15, in 1 series, lanceolate to elliptic, setose, up to 15 mm long; receptacle flat, epaleate; ligules up to 35 per head, bisexual, fertile, white or cream-colored, up to 15 mm long; cypselae oblong to linear, yellow-brown, 10- or 12-ribbed, 5–6 mm long, without a beak; pappus of numerous light brown capillary bristles 6–8 mm long, persistent.

Common Name: Great white lettuce; rattlesnake-root.
Habitat: Mesic woods, floodplain woods, banks of streams.
Range: New York to Minnesota, south to Arkansas, Kentucky, and West Virginia.
Illinois Distribution: Occasional throughout the state.

This species has the largest flowering heads of any species of *Nabalus* in Illinois. All of the leaves are petiolate, whereas the other species of *Nabalus* in Illinois have some leaves sessile.

Nabalus crepidineus flowers from August to October.

5. **Nabalus racemosus** (Michx.) Hook. Fl. Bor. Am. 1:294. 1833.
Prenanthes racemosa Michx. Fl. Bor. Am. 2:183. 1803.

Perennial herbs from a thickened taproot; latex present; stems erect, often unbranched, to 1.5 m tall, glabrous or sometimes with a few setae, glaucous; leaves basal and cauline, the basal and lower cauline leaves persistent, oblong to oblanceolate, obtuse at the apex, tapering to a winged petiole, glabrous, glaucous, the middle and upper cauline leaves similar but smaller and tapering to the sessile and sometimes clasping base; heads liguliform, numerous, in narrow racemes, to 6 mm across, erect or ascending, subtended by 8, usually purplish, setose, subulate bractlets up to 4 mm long; involucre campanulate, 4–7 mm high; phyllaries 7–14, in 1 series, lanceolate to linear, green or purple, setose, to 12 mm long; receptacle flat, epaleate; ligules 9–16 per head, bisexual, fertile, pinkish or purplish, to 12 mm long; cypselae more or less cylindrical, yellow-brown, several-ribbed, 5–6 mm long, without a beak; pappus of numerous pale yellow capillary bristles 6–7 mm long, persistent.

Nabalus racemosus is distinguished from the other species of *Nabalus* in Illinois by its setose phyllaries, narrow racemes, and pinkish or purplish flowering heads.

Two varieties occur in Illinois.
a. Phyllaries 7–10 per head; flowers 9–16 per head . . . . . . 5a. *N. racemosus* var. *racemosus*
a. Phyllaries 10–14 per head; flowers 17–26 per head . . . 5b. *N. racemosus* var. *multiflorus*

5a. **Nabalus racemosus** (Michx.) Hook. var. **racemosus**
Phyllaries 7–10 per head; flowers 9–16 per head.

Common Name: Rattlesnake-root.
Habitat: Prairies, moist soil.
Range: Newfoundland to British Columbia, south to Washington, Colorado, Missouri, and Kentucky.
Illinois Distribution: Occasional in the northern three-fourths of Illinois, less common southward.

5b. **Nabalus racemosus** (Michx.) Hook. var. **multiflorus** (Cronq.) Mohlenbr. Guide Ill. Fl., ed. 4, 138. 2013.
Prenanthus racemosa Michx. ssp. *multiflora* Cronq. Rhodora 50:30. 1848.

Phyllaries 10–14 per head; flowers 17–26 per head.

Common Name: Rattlesnake-root.
Habitat: Prairies, moist soil.
Range: Newfoundland to British Columbia, south to Washington, Colorado, Missouri, and Kentucky.
Illinois Distribution: Occasional in the northern one-third of Illinois.

This variety has more phyllaries and more flowers per head than var. *racemosus*. It has been found in the northern one-third of the state.

Variety *multiflorus* flowers from July to October.

110. **Sonchus** L.—Sow Thistle

Annual, biennial, or perennial herbs with taproots or rhizomes; latex present; stems erect, branched; leaves usually all cauline, alternate, usually pinnately divided, often spinulose-toothed; heads liguliform, borne in corymbs, not subtended by bractlets; involucre campanulate to ovoid; phyllaries numerous, in 3–5 series, unequal; receptacle flat, epaleate; ligules numerous, bisexual, fertile, usually yellow; cypselae oblong to linear-oblong, flattened, several-ribbed, without a beak; pappus of numerous white capillary bristles in up to 4 series, the inner connate at the base.

Sonchus consists of approximately 60 species native to Europe, Asia, and North Africa.

Four species occur in Illinois.
1. Flowering heads 3–5 cm across; flowers bright yellow.
  2. Phyllaries and peduncles stipitate-glandular .... 1. *S. arvensis*
  2. Phyllaries and peduncles eglandular or with sessile glands .... 2. *S. uliginosus*
1. Flowering heads 1.0–2.5 cm across; flowers pale yellow.
  3. Basal auricle of leaves rounded; cypselae longitudinally ribbed, otherwise smooth...
  .... 3. *S. asper*
  3. Basal auricle of leaves acute; cypselae longitudinally ribbed and papillate ....
  .... 4. *S. oleraceus*

1. **Sonchus arvensis** L. Sp. Pl. 2:793. 1753.
Perennial herbs from rhizomes; latex present; stems erect, branched, to 1.5 m tall, usually glabrous; leaves mostly cauline, alternate, the lowermost runcinate, acute at the apex, round-auriculate at the base, to 15 cm long, to 8 cm wide, spinulose-toothed, usually glabrous, the middle and upper cauline leaves lanceolate, less divided, sometimes entire; heads liguliform, borne in corymbs, 4–5 cm across, on stipitate-glandular peduncles, not subtended by bractlets; involucre campanulate to hemispheric, 15–25 mm high; phyllaries numerous, in 3–5 series,

dark green, without a scarious margin, stipitate-glandular, 14–17 mm long; receptacle flat, epaleate; ligules numerous per head, bisexual, fertile, yellow; cypselae narrowly oblong, flattened, about 10-ribbed, dark brown, 2.5–3.5 mm long; pappus of numerous white capillary bristles 8–14 mm long.

Common Name: Field sow thistle.
Habitat: Fields, disturbed soil.
Range: Native to Europe; introduced into most of North America.
Illinois Distribution: Occasional in the northern half of Illinois, uncommon in southern Illinois.

The species has the large showy flowering heads of *S. uliginosus* but differs in the stipitate glands on the nonscarious phyllaries and peduncles, the broadly campanulate and higher involucres, and the larger phyllaries. It is not as common as *S. uliginosus*.

Sonchus arvensis flowers from July to September.

2. **Sonchus uliginosus** Bieb. Fl. Taur. Caucas. 2:238. 1808.
Sonchus arvensis L. ssp. *uliginosus* (Bieb.) Nyman, Consp. Fl. Eur. 433. 1879.

Perennial herbs from rhizomes; latex present; stems erect, branched, to 1.5 m tall, usually glabrous; leaves mostly cauline, alternate, the lowermost runcinate, acute at the apex, round-auriculate at the base, to 15 cm long, to 8 cm wide, spinulose-toothed, usually glabrous, the middle and upper cauline leaves lanceolate, less divided, sometimes entire; heads liguliform, borne in corymbs on sessile-glandular peduncles, 4–5 cm across, not subtended by bractlets; involucre narrowly cylindric, 12–20 mm high; phyllaries numerous, in 3–5 series, pale green, with a conspicuous scarious margin, sessile-glandular, 10–15 mm long; receptacle flat, epaleate; ligules numerous per head, bisexual, fertile, yellow; cypselae oblong, flattened, 8- or 10-ribbed, brown, 2.5–3.5 mm long; pappus of numerous white capillary bristles to 15 mm long.

Common Name: Showy sow thistle.
Habitat: Fields, disturbed soil.
Range: Native to Europe and Asia; introduced throughout much of North America.
Illinois Distribution: Abundant in the northern half of Illinois, extending southward to Clay and St. Clair counties.

This showy weedy species is becoming increasingly abundant in the northern half of Illinois. Surprisingly, it has not been found in the southernmost counties of the state.

Some botanists consider this plant to be a subspecies of *S. arvensis*, a view not followed here. The following table summarizes some of the differences.

	S. arvensis	*S. uliginosus*
flowering head	on stipitate-glandular peduncles	on sessile-glandular peduncles
involucre	campanulate to hemispheric 15–25 mm high	narrowly cylindric 12–20 mm high
phyllaries	dark green; no scarious margins	pale green; conspicuous scarious margins
	stipitate-glandular; 14–17 mm long	sessile-glandular; 10–15 mm long
cypselae	oblong	narrowly oblong

Sonchus uliginosus flowers from July to September.

3. **Sonchus asper** (L.) Hill, Herb. Brit. 147. 1769.
Sonchus oleraceus L. var. *asper* L. Sp. Pl. 2:794. 1753.

Most annual herbs from a taproot; latex present; stems erect, branched or unbranched, up to 1.2 m tall, usually glabrous, hollow; leaves mostly cauline, the lowermost usually runcinate, acute at the apex, tapering to the petiolate base, the middle and upper cauline leaves pinnatifid or serrate, obovate to spatulate, acute at the apex, auriculate-clasping at the base, the auricles rounded, spiny-toothed, usually glabrous; heads liguliform, several to numerous, borne in corymbs, to 2.2 cm across, the peduncles usually glandular, not subtended by bractlets; involucre campanulate, to 13 mm high; phyllaries numerous, in 3–5 series, stipitate-glandular; receptacle flat, epaleate; ligules numerous per head, bisexual, fertile, yellow; cypselae oblong, flat, 3-ribbed on each face, glabrous, 2.5–3.5 mm long; pappus of numerous white capillary bristles up to 8 mm long.

Common Name: Spiny sow thistle.
Habitat: Fields, disturbed areas.
Range: Native to Europe, Asia, and North Africa; introduced throughout North America.
Illinois Distribution: Occasional to common throughout the state.

This common weed of disturbed areas is similar in appearance to *S. oleraceus* but differs in the rounded auricles of the leaf bases and its glabrous cypselae.
Sonchus asper flowers from June to October.

4. **Sonchus oleraceus** L. Sp. Pl. 2:794. 1753.
Mostly annual herbs from a taproot; latex present; stems erect, branched or unbranched, to 1.5 m tall, usually glabrous, hollow; leaves mostly cauline, the lowermost usually runcinate and petiolate, the middle and upper ones pinnatifid or serrate, with spine-tipped teeth, obovate to spatulate, acute at the apex, tapering to

the auriculate-clasping base, the auricles acute, usually glabrous; heads liguliform, several to numerous, borne in corymbs, to 2.2 cm across, the peduncles stipitate-glandular or glabrous, not subtended by bractlets; involucre campanulate, to 13 mm high; phyllaries numerous, in 3–5 series, usually glabrous, less commonly stipitate-glandular; receptacle flat, epaleate; ligules numerous per head, bisexual, fertile, yellow; cypselae oblong, not noticeably flattened, brown, 2.5–3.5 mm long, 4-nerved on each face, papillate; pappus of numerous white capillary bristles up to 8 mm long.

Common Name: Common sow thistle.
Habitat: Fields and disturbed areas.
Range: Native to Europe; introduced into all of North America.
Illinois Distribution: Common throughout the state; probably in every county.

This is the most widespread species of *Sonchus* in Illinois. It may be distinguished from the very similar-appearing *S. asper* by its acute auricles at the base of the leaves and by its papillate cypselae.

Sonchus oleraceus flowers from June to October.

III. **Hieracium** L.—Hawkweed

Perennial herbs from taproots or stolons; latex present; stems erect, branched, sometimes scapose; leaves basal and/or cauline and alternate, entire to dentate to pinnatifid; heads liguliform, solitary or in various types of inflorescences, sometimes subtended by bractlets; involucre campanulate to hemispheric to cylindrical; phyllaries up to 40, in 2 series, sometimes unequal, usually without a scarious margin; receptacle flat, epaleate, pitted; ligules up to 150 per head, bisexual, fertile, usually yellow or orange; cypselae usually columnar, dark red to black, 10-ribbed, without a beak; pappus of several to many white or sordid barbellate capillary bristles in 1–2 series.

There may be as many as 1,000 species in the genus, native to almost all parts of the world except Australia and Antarctica. It is the only genus in tribe Cichorieae that does not have runcinate or pinnatifid leaves.

Eight species have been found in Illinois, half of them native.

1. Plants with well-developed clusters of basal leaves at flowering time.
   2. Basal leaves rounded to cordate at the base; pappus white . . . . 8. *H. murorum*
   2. Leaves tapering to the base; pappus sordid.
      3. Plants without stolons . . . . 3. *H. piloselloides*
      3. Plants with stolons.
         4. Ligules red-orange. . . . . .1. *H. aurantiacum*
         4. Ligules yellow. . . . . 2. *H. caespitosum*
1. Plants without well-developed clusters of basal leaves at flowering time.
   5. Leaves more than 24 per stem . . . . 6. *H. umbellatum*
   5. Leaves less than 24 per stem.
      6. At least some of the pubescence more than 1 cm long . . . . 4. *H. longipilum*
      6. None of the pubescence 1 cm long.
         7. Heads with more than 40 ligules; cypselae truncate at summit . . . 7. *H. scabrum*
         7. Heads with less than 40 ligules; cypselae tapering to summit . . . . 5. *H. gronovii*

1. **Hieracium aurantiacum** L. Sp. Pl. 2:801. 1753.
Pilosella aurantiaca (L.) F. W. Schultz-Bip. Flora 45:426. 1852.

Perennial herbs with stolons; latex present; stems scapose, rarely with 1 or 2 leaves, unbranched, to 75 cm tall, hirsute; leaves mostly basal, in rosettes, spatulate to oblong, obtuse or acute at the apex, tapering to the usually sessile base, to 10 cm long, to 4 cm wide, entire, hirsute on both surfaces and with some stellate pubescence; heads liguliform, several, borne in corymbs, up to 2 cm across, the peduncles stipitate-glandular, subtended by up to 8 bractlets; involucre campanulate; phyllaries up to 30, in 2 series, somewhat unequal, hirsute and stipitate-glandular; receptacle flat, epaleate, ribbed; ligules numerous per head, bisexual, fertile, red-orange to orange, up to 14 mm long; cypselae oblongoid, dark red, 1.2–2.0 mm long, glabrous or nearly so; pappus of up to 30 brown capillary bristles in 1 series, 3.5–4.0 mm long.

Common Name: Orange hawkweed; devil's paintbrush.
Habitat: Adventive in grassy areas and sandy fields, woods, waste places.
Range: Native to Europe; introduced into several U.S. states.
Illinois Distribution: Occasional in the northeastern counties.

This is the only species of *Hieracium* in Illinois with orange flowering heads. Species of *Hieracium* that have stolons and/or rhizomes rather than taproots have sometimes been placed in the genus *Pilosella*.

Hieracium aurantiacum flowers from July to September.

2. **Hieracium caespitosum** Dum. Fl. Belg. 62. 1827.
Hieracium pratense Tausch. Flora II, part 1, Erg. 56. 1828.
Pilosella caespitosa (Dum.) P. D. Sell & C. West, Watsonia 6:314. 1861.

Perennial herbs with stolons and rhizomes; latex present; stems erect, scapose, unbranched, to 80 cm tall, hirsute and stipitate-glandular; leaves basal, in rosettes, with up to 3 cauline leaves, oblong to oblanceolate to lanceolate, obtuse or acute at the apex, tapering to the sessile base, to 25 cm long, to 10 cm wide, entire or denticulate, hirsute and stipitate-glandular; heads liguliform, up to 25, borne in umbels, up to 2 cm across, on hirsute, glandular-stipitate peduncles, subtended by up to 8 bractlets; involucre campanulate; phyllaries up to 18, in 1 series, somewhat unequal, hirsute and stipitate-glandular; receptacle flat, epaleate, pitted; ligules up to 50 per head, bisexual, fertile, yellow; cypselae columnar, dark red, 1.5–1.8 mm long, truncate at the summit, glabrous or nearly so; pappus of up to 30 white capillary bristles in series, 4–6 mm long.

Common Name: King devil; field hawkweed.
Habitat: Disturbed areas, oak woods, savannas, prairies.
Range: Native to Europe; introduced into several U.S. states and Canada.
Illinois Distribution: Occasional in the northeastern counties.

This species is similar in many characteristics to *H. aurantiacum*, differing in its yellow flowering heads rather than orange.

For a number of years, this species was known as *H. pratense*, but that binomial was preceded by *H. caespitosum* by 1 year.

Hieracium caespitosum flowers from May to August.

3. **Hieracium piloselloides** Vill. Prosp. Hist. Pl. Dauphine 34. 1779.
Hieracium florentinum All. Fl. Ped. 1:213. 1785.
Pilosella piloselloides (Vill.) Sojak, Preslia 43:185. 1971.

Perennial herbs from a short thick rhizome, but lacking stolons; latex present; stems scapose, unbranched, to nearly 1 m tall, glabrous or with a few hirsute hairs, glaucous; leaves mostly basal, in rosettes, with a few reduced cauline leaves, oblong to spatulate, obtuse to acute at the apex, tapering to the sessile base, entire, glabrous or hirsute on the margins and the veins on the lower surface; heads liguliform, up to 30, borne in corymbs, 10–14 mm across, the peduncles hirsute and glandular, subtended by up to 12 bractlets; involucre campanulate; phyllaries up to 18, in 1 series, usually unequal, hirsute and stipitate-glandular; receptacle flat, epaleate, pitted; ligules numerous, bisexual, fertile, yellow; cypselae oblongoid, 1.5–2.0 mm long, truncate at the summit, usually glabrous; pappus of up to 40 white or brownish capillary bristles in 1 series, 3–4 mm long.

Common Name: Glaucous king devil.
Habitat: Disturbed soil.
Range: Native to Europe; adventive in several U.S. states and Canada.
Illinois Distribution: Known only from Lake County.

This is the only member of the scapose species of *Hieracium* that lacks stolons. The stems are usually somewhat glaucous.

For a number of years, this species was known as *H. florentinum*.

Hieracium piloselloides flowers during May and June.

4. **Hieracium longipilum** Torr. ex Hook. Fl. Bor. Am. 1:298. 1833.
Perennial herbs from a taproot; latex present; stems erect, branched, to 1.5 m tall, with usually rusty hairs 10–20 mm long; leaves basal and cauline, with several cauline leaves crowded near the base of the stem, oblanceolate, obtuse to acute at the apex, tapering to the usually sessile base, to 10 cm long, to 3.5 cm wide, entire, densely covered with long pubescence, the hairs 10–20 mm long, the middle and upper leaves up to 12 in number, smaller; heads liguliform, up to 20 per plant, up to 18 mm across, borne mostly in racemes, the peduncles with long hairs, stipitate-glandular, subtended by up to 12 bractlets; involucre campanulate; phyllaries up to 20, in 1 series, equal, stipitate-glandular; receptacle flat, epaleate; ligules numerous, bisexual, fertile, yellow; cypselae fusiform, red-brown, 3–4 mm long; pappus of numerous stramineous or sordid capillary bristles in 2 series, 5.0–6.5 mm long.

Common Name: Long-bearded hawkweed.
Habitat: Fields, prairies, open woods, black oak savannas.
Range: Ontario to Minnesota, south to Texas, Louisiana, and Tennessee.
Illinois Distribution: Occasional throughout the state.

The pubescence of the stems and leaves is 10 mm long or longer, much longer than in any other species of *Hieracium* in Illinois.

    Hieracium longipilum flowers from July to September.

5. **Hieracium gronovii** L. Sp. Pl. 2:801. 1753.
    Perennial herbs from a taproot; latex present; stems erect, branched, to 90 cm tall, hirsute with papillate-based hairs less than 10 mm long, stipitate-glandular; leaves mostly cauline, although usually several of them crowded toward the base of the plant, up to 12 in number, obovate to oblanceolate, obtuse to acute at the apex, tapering to the usually sessile base, to 8 cm long, to 4 cm wide, entire with hairs up to 10 mm long covering both surfaces; heads liguliform, up to 40 in a thyrse or raceme, up to 14 mm across, the peduncles hirsute and stipitate-glandular, subtended by up to 12 bractlets; involucre cylindric or campanulate; phyllaries up to 15 in 1 series, glabrous or stellate-pubescent, but usually not stipitate-glandular; receptacle flat, epaleate, pitted; ligules up to 20 per head, bisexual, fertile, yellow; cypselae fusiform, tapering to the apex, red-brown, 3.5–4.5 mm long; pappus of numerous stramineous capillary bristles in 2 series, 4.5–5.5 mm long.

Common Name: Hairy hawkweed.
Habitat: Dry open woods, black oak savannas, oak barrens, prairies.
Range: Maine to Minnesota, south to Texas and Florida; Ontario.
Illinois Distribution: Occasional in Illinois, although rare or absent in the
    northwestern counties.

This species is similar to *H. longipilum* but has much shorter hairs on the stems and leaves. It is also similar to *H. scabrum* but has fewer flowers per head and cypselae tapering to the summit. *Hieracium umbellatum* is also similar, but this species has more than 24 leaves on the stem.

    Hieracium gronovii flowers from June to October.

6. **Hieracium umbellatum** L. Sp. Pl. 2:804. 1753.
Hieracium kalmii L. Sp. Pl. 2:804. 1753, misapplied.
Hieracium canadense Michx. Fl. Bor. Am. 2:86. 1803.
Hieracium fasciculatum Pursh, Fl. Am. Sept. 2:504. 1813.
Hieracium scabriusculum Schwein. in Long's Exp. 2:394. 1824.
Hieracium canadense Michx. var. *fasciculatum* (Pursh) Fern. Rhodora 45:320. 1943.
Hieracium kalmii L. var. *fasciculatum* (Pursh) Lepage, Nat. Can. 87:87. 1960.
Hieracium kalmii L. var. *canadense* (Michx.) Reveal, Novon 3:73. 1993.

Perennial herbs from a taproot; latex present; stems erect, branched, to 1.5 m tall, glabrous above, hirsute and usually stellate-pubescent below; leaves mostly cauline, at least 24 per stem, lanceolate to narrowly oblong, obtuse to acute at the apex, tapering to the sessile and sometimes clasping base, to 8 cm long, to 2.5 cm wide, denticulate or sometimes entire, usually glabrous on both surfaces or occasionally scabrous or stellate-pubescent; heads liguliform, numerous, borne in corymbs or umbels, to 18 mm across, the peduncles stellate-pubescent, subtended by up to 15 bractlets; involucre campanulate; phyllaries up to 20, in 2–3 series, usually glabrous, more or less equal; receptacle flat, epaleate, pitted; ligules numerous per head, bisexual, fertile, yellow; cypselae columnar, truncate at the summit, 2.5–3.5 mm long, red-brown; pappus of numerous stramineous to sordid capillary bristles in 2 series, 6–7 mm long.

Common Name: Canada hawkweed.
Habitat: Dry woods, sand barrens.
Range: Newfoundland to Alaska, south to Oregon, Colorado, Missouri, and West Virginia.
Illinois Distribution: Occasional in the northern one-sixth of the state, absent elsewhere.

This species differs from all other species of *Hieracium* in Illinois in its many leaves on the stem. The phyllaries are in 2–3 series, as are the pappi. This is the only species in *Hieracium* with regularly denticulate leaves.

There has been considerable confusion as to the correct binomial for this species. For years it was generally called *H. canadense* and sometimes *H. kalmii*. Linnaeus's *H. kalmii* was misapplied to this species, and *H. canadense* is predated by *H. umbellatum*. Our plants in the past were usually called *H. canadense* var. *fasciculatum*.

Hieracium umbellatum flowers during August and September.

7. **Hieracium scabrum** Michx. Fl. Bor. Am. 2:86. 1803.
Perennial herbs from a taproot; latex present; stems erect, branched, to 1.5 m tall, hirsute or sometimes glabrous or nearly so above; leaves mostly cauline, up to 20 in number, obovate to oblanceolate, acute or obtuse at the apex, tapering to the usually sessile base, to 10 cm long, to 4 cm wide, entire, hirsute and scabrous on both surfaces; heads liguliform, numerous, to 15 mm across, borne in corymbs, the peduncles stipitate-glandular, subtended by up to 15 bractlets; involucre campanulate; phyllaries up to 20, in 2–3 series, stellate-pubescent and stipitate-glandular, more or less equal; receptacle flat, epaleate, pitted; ligules numerous per head, bisexual, fertile, yellow; cypselae columnar, truncate at the apex, red-brown, 2.5–3.0 mm long; pappus of numerous stramineous capillary bristles in 2 series, 6–7 mm long.

This species differs from *H. umbellatum* in having fewer cauline leaves, from *H. longipilum* in having much shorter pubescence, and from *H. gronovii* in having more flowers per head and in its truncate-tipped cypselae.

Two varieties may be recognized in Illinois.
a. Pubescence on the stem up to 3 mm long . . . . .7a. *H. scabrum* var. *scabrum*
a. Pubescence on the stem 3–10 mm long . . . . 7b. *H. scabrum* var. *intonsum*

7a. **Hieracium scabrum** Michx. var. **scabrum**
Pubescence on the stem up to 3 mm long.

Common Name: Rough hawkweed.
Habitat: Black oak savannas, fields, dry woods, prairies.
Range: Quebec to Ontario to Minnesota, south to Oklahoma, Arkansas, and Georgia.
Illinois Distribution: Occasional throughout Illinois.

This is the common variety of *H. scabrum* in Illinois.

A hybrid between this variety and *H. umbellatum*, called *Hieracium X fassettii* Lepage, has been reported from Illinois.

Hieracium scabrum var. *scabrum* flowers during August and September.

7b. **Hieracium scabrum** Michx. var. **intonsum** Fern. & St. John, Rhodora 16:183. 1914.
Pubescence on the stem 3–10 mm long.

Common Name: Rough hawkweed.
Habitat: Black oak savannas, fields, dry woods, prairies.
Range: Illinois, Iowa, and Missouri.
Illinois Distribution: Uncommon in Illinois.

This variety, questionably distinct from the typical variety, has been found a few times in the state. It flowers during August and September.

8. **Hieracium murorum** L. Sp. Pl. 2:802. 1753.
Perennial herbs with a short rhizome; latex present; stems erect, usually unbranched, up to 75 cm tall, hirsute, stipitate-glandular, stellate-pubescent; leaves mostly basal, with up to 3 cauline leaves, the basal leaves elliptic to oblong, obtuse to acute at the apex, cordate or truncate or rounded at the petiolate base, to 8 cm long, to 4 cm wide, usually purple-mottled, toothed, at least near the base, hirsute and scabrous on both surfaces; heads liguliform, few to several, borne in corymbs, to 35 mm across, the peduncles stellate-pubescent and stipitate-glandular, subtended by up to 15 bractlets; involucre more or less campanulate; phyllaries up to 20, in 2–3 series, unequal, stipitate-glandular, stellate-pubescent; receptacle flat, epaleate, pitted; ligules numerous per head, bisexual, fertile, yellow; cypselae columnar, truncate at the apex, red-brown, 2.5–3.0 mm long; pappus of numerous stramineous capillary bristles in 2 series, 4–5 mm long.

Common Name: Golden lungwort.
Habitat: Disturbed soil.
Range: Native to Europe; introduced into a few midwestern and northeastern states.
Illinois Distribution: Known only from Sangamon County.

Because of the sometimes purple-mottled leaves, this species resembles *H. venosum*, but it is not as strongly purple. Although *Hieracium venosum* L. was reported from Illinois in *Flora of North America*, John Strother (personal correspondence) has checked the specimen on which this report was based and found it to be *H. gronovii*.

Hieracium murorum flowers from June to August.

112. Leontodon L.—Hawkbit

Annual or perennial herbs from fibrous roots or a taproot; latex present; stems scapose, unbranched, with or without scales; leaves all in basal tufts, some or all of them pinnatifid, glabrous or pubescent; head liguliform, solitary, showy, the peduncle subtended by up to 20 bractlets; involucre campanulate; phyllaries up to 20, in 2 series, nearly equal, glabrous or pubescent; receptacle convex, epaleate, pitted, villous; ligules up to 30 per head, bisexual, fertile, yellow; cypselae oblongoid to cylindric, in 2 series, brown, curved, usually without a beak, several-ribbed, muricate, glabrous; pappus sometimes of two types, that of the outer cypselae reduced to scales, that of the inner cypselae of brownish plumelike bristles, or pappus only of plumose bristles.

This genus comprises about 50 species, native to Europe, Asia, and North Africa.

Species of *Leontodon* resemble *Taraxacum*, differing in their plumose or scalelike pappus and the absence of a beak on the cypselae.

Two introduced species occur in Illinois.

1. Pappus of all flowers with a single row of plumose bristles . . . . 1. *L. autumnalis*
1. Pappus of 2 types, the inner double with plumose and setiform bristles, the outer reduced to a short irregular crown . . . . 2. *L. saxatilis*

1. **Leontodon autumnalis** L. Sp. Pl. 2:798. 1753.
Apargia autumnalis (L.) Hoffm. Deutsch. Fl., ed. 2, 2:113. 1800.
Oporinia autumnalis (L.) D. Don, Edin. New Philos. 6:309. 1829.

Perennial herbs with a taproot; latex present; stems scapose, but with a few bractlets, usually branched, to 80 cm tall, glabrous or tomentose near the summit; leaves all basal, linear-oblanceolate, pinnatifid or sometimes laciniate, obtuse to acute at the apex, tapering to the base, to 25 cm long, to 3 cm wide, glabrous or less commonly hirsute on both surfaces; heads liguliform, solitary or usually few in a corymb, 2.0–3.5 cm across, subtended by 15–20 subulate bractlets up to 4 mm long; involucre campanulate; phyllaries up to 20, in 2 series, narrowly lanceolate, slightly unequal, glabrous or pubescent, to 12 mm long; receptacle convex, epaleate, pitted, villous; ligules up to 30 per head, 12–16 mm long, bisexual, fertile, yellow; cypselae cylindric or fusiform, brown, curved, 4–7 mm long, without a beak; pappus a single row of tan plumose bristles 5–8 mm long.

Common Name: Fall dandelion.
Habitat: Disturbed soil.
Range: Native to Europe and Asia; introduced into a few U.S. states.
Illinois Distribution: Known only from Champaign and Christian counties.

This species looks a lot like species of *Taraxacum* but usually has more than one head per plant and a beakless cypsela.

Leontodon autumnalis flowers from June to September.

2. **Leontodon saxatilis** Lam. Fl. Franc. 2:115. 1779.
Thrincia leysseri Sched. Crit. 44. 1822.
Leontodon taraxicoides (Villars) Willd. ex Merat. Ann. Sci. Nat. Paris 22:108. 1831, *non* Hoppe & Horns. (1821).
Leontodon leysseri (Sched.) G. Beck, Fl. Nieder-Ost. 2:1312. 1893.

Perennial herbs with a taproot; latex present; stems scapose, erect to ascending, unbranched, without bractlets, to 30 cm tall, glabrous or hispid; leaves all basal, narrowly oblong, obtuse to acute at the apex, tapering to the base, to 12 cm long, to 2.5 cm wide, pinnatifid or less commonly entire or toothed, hispid on both surfaces; head liguliform, solitary, 2–3 cm across, the peduncle subtended by up to 16 subulate bractlets up to 4 mm long; involucre campanulate; receptacle convex. epaleate, pitted, villous; ligules up to 30 per head, 10–15 mm long, bisexual, fertile, yellow; cypselae fusiform, in 1–2 series, curved, tan, 4–6 mm long, those in the outer series without a beak, those of the inner series with a short beak; pappus in 2 series, the outer of minute scales, the inner of pale tan plumose bristles 5–6 mm long.

Common Name: Hawkbit.
Habitat: Disturbed soil, particularly in lawns and cemeteries.
Range: Native to Europe; introduced into several U.S. states.
Illinois Distribution: Known only from Cook and DuPage counties.

This species differs from *L. autumnalis* and species of *Taraxacum* in having two types of pappi.

Our plants in the past have been known as *L. taraxicoides*, which is an illegitimate binomial, and *L. leysseri*, which is predated by *L. saxatilis*.

Leontodon saxatilis flowers from June to September.

113. **Hypochaeris** L.—Cat's-ear

Annual or perennial herbs with a taproot; latex present; stems erect, branched or unbranched, usually scapose but with a few scales; leaves basal, glabrous or pubescent, entire to pinnatifid; heads liguliform, solitary or few in cymes or corymbs, the peduncles with scales, not subtended by bractlets; involucre cylindric or campanulate; phyllaries up to 30, in 3–4 series, unequal, glabrous or pubescent; receptacle flat, paleate, pitted; ligules numerous per head, bisexual, fertile, yellow; cypselae in 2 series, the outer beaked or beakless, the inner beaked, brown, up to 10-ribbed, often muricate; pappus in 1 or 2 series: if in 1 series, composed of white or tan plumose bristles; if in 2 series, the outer of barbellate capillary bristles, the inner of shorter plumose bristles.

There are about 60 species in this genus, native to Europe, Asia, North Africa, and South America. The genus is sometimes spelled *Hypochoeris*.

Hypochaeris is very similar in appearance to *Leontodon* but differs in its paleate receptacle and its unequal phyllaries.

Two species have been found in Illinois.

1. Leaves glabrous or pubescent only on the midrib; outer cypselae beakless, inner cypselae beaked; annuals . . . . 1. *H. glabra*
1. Leaves hispid; all cypselae beaked; perennials . . . . 2. *H. radicata*

1. **Hypochaeris glabra** L. Sp. Pl. 2:811. 1753.

Annual herbs from a taproot; latex present; stems erect, usually branched, nearly scapose, to 30 cm tall, with a few scales, glabrous; leaves nearly all basal, narrowly oblong to oblanceolate, acute to obtuse at the apex, tapering to the sessile base, to 10 cm long, to 2.5 cm wide, usually pinnatifid, glabrous except for the midrib; heads liguliform, 1–3 in a cyme, not subtended by bractlets; involucre narrowly campanulate; phyllaries up to 20, in 3–4 series, unequal, up to 18 mm long, with scarious margins, glabrous, sometimes ciliolate; receptacle flat, paleate, pitted; ligules up to 40 per head, up to 8 mm long, bisexual, fertile, yellow; cypselae of 2 types, the outer cylindric, truncate at the summit, the inner fusiform, 3–4, with a slender beak 3–4 mm long, both types dark brown, 10-ribbed, 8–10 mm long; pappus in 2 series, 9–10 mm long, tawny, the outer of barbellate capillary bristles, the inner of plumose bristles.

Common Name: Smooth cat's-ear.
Habitat: Disturbed soil.
Range: Native to Europe; introduced into several U.S. states, particularly in the Southeast.
Illinois Distribution: Known only from Jackson County, where it was collected by Paul Thomson in the 1970s.

This rather inconspicuously flowering weed differs from *H. radicata* in its usually glabrous herbage, its outer row of beakless cypselae, and its annual habit.

Hypochaeris glabra flowers from May to August.

2. **Hypochaeris radicata** L. Sp. Pl. 2:811. 1753.

Perennial herbs from long taproots; latex present; stems erect, usually scapose but sometimes with a few scales, up to 50 cm tall, hirsute; leaves all basal, oblanceolate to obovate, obtuse to acute at the apex, tapering to the sessile or short-petiolate base, to 30 cm long, to 3 cm wide, deeply pinnatifid, hirsute; heads liguliform, few in a corymb, the peduncles not subtended by bractlets; involucre narrowly campanulate; phyllaries up to 30, in 3–4 series, unequal, to 20 mm long, with scarious margins, glabrous or sparsely hirsute; receptacle flat, paleate, pitted; ligules up to 40 per head, to 15 mm long, bisexual, fertile, yellow; cypselae all alike, fusiform, beaked, brown, with 10–12 ribs, muricate, 6–10 mm long, the beak 3–5 mm long; pappus in 2 series, 10–12 mm long, white, the outer of barbellate capillary bristles, the inner of plumose bristles.

Common Name: Rough cat's-ear.
Habitat: Disturbed soil, particularly in old fields, lawns, and cemeteries.
Range: Native to Europe and Asia; introduced into several U.S. states.
Illinois Distribution: Known from Champaign, Cook, DuPage, Kane, McHenry, and
St. Clair counties.

This perennial species has all cypselae beaked. It also has hirsute herbage, longer
ligules, and pappus bristles longer than those of the annual *H. glabra*.

Hypochaeris radicata flowers from May to August.

114. **Helminotheca** Zinn.—Ox-tongue

Annual or biennial herbs with taproots; latex present; stems erect, branched,
leafy, hispid or hirsute; leaves basal and cauline, some of them usually pinnatifid,
pubescent, the basal petiolate, the cauline sessile; heads liguliform, 1 to several in
corymbs, the peduncles without scales but subtended by up to 5 foliaceous bractlets; involucre campanulate or urceolate; phyllaries up to 12, in 1 series, subequal,
seldom with a scarious margin; receptacle flat, epaleate, pitted, glabrous; ligules
numerous per head, bisexual, fertile, yellow; cypselae in 2 series, short-beaked, the
outer 5- to 10-ribbed, pubescent, the inner rugulose to muricate, glabrous; pappus
up to 15, in 2 series, white, with barbellate or plumose bristles or subulate scales.

There are 4 species native to Europe. At one time they were included in *Picris*.
They differ from *Picris* in having shorter beaks on the cypselae and foliaceous
bractlets.

Only the following species is known from Illinois.

1. **Helminotheca echioides** (L.) Holub, Folia Geobot. Phytotax. 8:176. 1973.
Picris echioides L. Sp. Pl. 2;792. 1753.

Annual or biennial herbs with a taproot; latex present; stems erect, branched,
leafy, to 75 cm tall, setose, hispid; leaves basal and cauline, the basal lanceolate to
spatulate, obtuse to acute at the apex, tapering to the short-petiolate base, to 15 cm
long, to 8 cm wide, usually some of them pinnatifid, hirsute, the cauline smaller,
not pinnatifid, clasping or sessile at the base, hirsute; heads liguliform, 1–5 in a
corymb, the peduncles hispid, subtended by 5 foliaceous bractlets up to 15 mm
long; involucre urceolate; phyllaries up to 12, in 1 series, subequal, without scarious margins, bristly at the apex; receptacle flat, epaleate, pitted, glabrous; ligules
up to 60 per head, 5–10 mm long, bisexual, fertile; cypselae in 2 series, fusiform,
brown, short-beaked, the outer with 5–10 ribs, pubescent, the inner rugulose to
muricate, glabrous, both types 2.5–3.0 mm long, the beaks 2.5–3.0 mm long; pappus up to 15, in 2 series, the bristles plumose, 4–7 mm long.

Common Name: Bristly ox-tongue.
Habitat: Disturbed soil.
Range: Native to Europe; introduced into a few U.S. states.
Illinois Distribution: Known only from Hancock County.

This species is distinguished by its leafy stems, its short-beaked cypselae, and its plumose pappi.

Helminotheca echioides flowers from July to September.

115. **Picris** L.—Bitterweed

Biennial or perennial herbs (in Illinois) or annuals, with a taproot, fibrous roots, or rhizomes; latex present; stems erect, branched, pubescent; leaves basal and cauline, the basal usually absent at flowering time, the cauline alternate, entire to toothed to pinnatifid, pubescent; heads liguliform, borne in corymbs, the peduncles sometimes with scales, subtended by several bractlets; involucre campanulate to urceolate; phyllaries up to 12, in 1 series, equal, with scarious margins, reflexed at fruiting; receptacle flat or convex, epaleate, pitted, glabrous; ligules numerous per head, bisexual, fertile, yellow; cypselae oblongoid or fusiform, beaked, ribbed, rugulose, glabrous; pappus of numerous white or stramineous barbellate or plumose bristles in 2–3 series.

This genus comprises about 40 species. Species that used to be in *Picris* that have short beaks on the cypselae and foliaceous bractlets have now been transferred to *Helminotheca*.

Only the following species occurs in Illinois.

1. **Picris hieracioides** L. Sp. Pl. 2:792. 1753.

Biennial or perennial herbs with a taproot; latex present; stems erect, leafy, branched, to 1 m tall, setose, hirsute, or hispid; leaves mostly cauline at flowering time, lanceolate to oblong-lanceolate, obtuse to acute at the apex, tapering to the sessile or petiolate base, up to 15 cm long, up to 7 cm wide, entire to dentate, hispid or hirsute; heads liguliform, numerous, to 10 mm across, the peduncle with a few scales, subtended by several narrowly lanceolate bractlets; involucre urceolate; phyllaries up to 12, in 1 series, lanceolate, bristly or tomentose; receptacle flat or convex, epaleate, pitted, glabrous; ligules numerous per head, bisexual, fertile, yellow; cypselae oblongoid, red-brown, 3–5 mm long, with a slender beak, rugulose, glabrous; pappus of numerous white plumose bristles 5–7 mm long, in 2–3 series.

Common Name: Cat's-ear.
Habitat: Disturbed soil.
Range: Native to Europe and Asia; introduced into a few eastern states.
Illinois Distribution: Known only from Menard County.

This rare introduction has not been seen in Illinois for many years. It differs from the similar appearing *Helminotheca echioides* in its slender beak of the cypselae and the absence of foliaceous bractlets.

Picris hieracioides flowers from July to September.

116. **Tragopogon** L.—Goatsbeard

Mostly biennial herbs with a taproot; latex present; stems erect, branched, glabrous or pubescent, not setose; leaves basal and cauline, alternate, usually entire, not lobed, glabrous or pubescent, often clasping; head liguliform, borne singly, the peduncles without scales, not subtended by bractlets; involucre campanulate; phyllaries up to 15, in 1 series, equal, with scarious margins, glabrous or pubescent; receptacle convex, epaleate, not pitted, glabrous; ligules numerous per head, bisexual, fertile, purple or yellow; cypselae cylindric to fusiform, usually beaked, brown or stramineous, several-ribbed, the ribs often muricate; pappus of up to 20 white or brownish, basally connate, plumose bristles in 1 series.

There may be as many as 150 species in this genus, native to Europe, Asia, and North Africa. Three species and 1 hybrid have been found in Illinois.

1. Flowers yellow.
- 2. Phyllaries equaling or shorter than the flowering heads; peduncle not or scarcely enlarged below the flowering head . . . . 1. *T. pratensis*
   2. Phyllaries longer than the flowering heads; peduncle enlarged below the flowering head . . . . 2. *T. dubius*
1. Flowers purple . . . . 3. *T. porrifolius*

1. **Tragopogon pratensis** L. Sp. Pl. 2:789. 1753.

Biennial herbs with a taproot; latex present; stems erect, branched, to 1 m tall, usually glabrous; leaves mostly cauline, alternate, broadly lanceolate, acuminate and recurved at the apex, clasping at the base, keeled, to 8 cm long, to 2.5 cm wide, entire, usually glabrous, at least at maturity; head liguliform, solitary, to 4 cm across, on peduncles not enlarged beneath the flowering head at anthesis, not subtended by bractlets; involucre urceolate; phyllaries usually 8, in 1 series, equaling or shorter than the flowering head, equal, with a scarious margin, usually glabrous; receptacle convex, epaleate, not pitted, glabrous; ligules numerous per head, bisexual, fertile, yellow, the outer often longer than the phyllaries; cypselae fusiform, brown, 5–6 cm long (including the long beak), 5- to 10-ribbed, the ribs muricate; pappus of numerous white capillary plumose bristles 3–4 cm long, forming a fruiting head up to 6 cm across.

Common Name: Common goatsbeard.
Habitat: Fields, along railroads, other disturbed soil.
Range: Native to Europe; introduced into most of the United States.
Illinois Distribution: Common throughout the state.

This species is similar to *T. dubius* because of its yellow flowering heads but differs in the peduncle not being enlarged below the flowering head at anthesis, the leaves having recurved tips, and the outer ligules usually being longer than the phyllaries.

Tragopogon pratensis flowers from May to November.

2. **Tragopogon dubius** Scop. Fl. Carniol., ed. 2, 2:95. 1772.
Tragopogon major Jacq. Fl. Austriaca 1:19. 1773.

Biennial herbs with a taproot; latex present; stems erect, branched, to 1 m tall, glabrous; leaves mostly cauline, alternate, keeled, broadly lanceolate, acuminate but not recurved at the tip, clasping as the base, to 8 cm long, to 2.5 cm wide, entire, glabrous at maturity; head liguliform, solitary, to 4 cm across, on peduncles conspicuously enlarged below the flowering head at anthesis, not subtended by bractlets; involucre conic; phyllaries usually 8, in 1 series, equal, with a scarious margin, longer than the flowering head, glabrous; ligules numerous per head, bisexual, fertile, yellow, all of them shorter than the phyllaries; cypselae narrowly fusiform, brown, 2–4 cm long (including the long beak), 5- to 10-ribbed, the ribs muricate; pappus of numerous white capillary plumose bristles 2–3 cm long, forming a fruiting head up to 5 cm across.

Common Name: Sand goatsbeard.
Habitat: Fields and disturbed soil.
Range: Native to Europe, North Africa, and Australia; introduced into most of the United States and Canada.
Illinois Distribution: Common throughout the state.

This species is readily distinguished from *T. pratensis* by its ligules being much shorter than the phyllaries and by the enlarged peduncle just below the flowering head.
   Tragopogon dubius flowers from May to September.

3. **Tragopogon porrifolius** L. Sp. Pl. 2:789. 1753.
   Biennial herbs with a taproot; latex present; stems erect, branched, to 1 m tall, glabrous; leaves mostly cauline, alternate, keeled, broadly lanceolate, acuminate but not recurved at its apex, clasping at the base, to 8 cm long, to 2.5 cm wide, entire, glabrous; head liguliform, solitary, to 8 cm across, the peduncle enlarged below the flowering head at anthesis, not subtended by bractlets; involucre conic; phyllaries up to 12 (–15), in 1 series, equal, with a scarious margin, glabrous; receptacle conic, epaleate, pitted, glabrous; ligules numerous per head, all of them shorter than the phyllaries, bisexual, fertile, purple; cypselae broadly fusiform, brown, 5–6 cm long (including the long beak), 5- to 10-ribbed, the ribs muricate; pappus of numerous white capillary plumose bristles 5–6 cm long, forming a fruiting head up to 6 cm across.

Common Name: Salsify; vegetable oyster.
Habitat: Fields, disturbed soil.
Range: Native to Europe and North Africa; introduced into most of the United States and Canada.
Illinois Distribution: Occasional throughout Illinois.

The purple flowering heads of this species are distinctive.
   A presumed hybrid between *T. porrifolius* and *T. pratensis*, known as *Tragopogon* X *neohybridus* Farw., has been found in Kane County. It has flowering heads that are more or less red.
   Tragopogon porrifolius flowers during May and June.

117. **Nothocalais** (Gray) Greene—Prairie Dandelion

Perennial herbs with a taproot; latex present; stems 1 (in Illinois) to several, erect, scapose, glabrous or pubescent; leaves basal, petiolate or sessile, entire, glabrous or pubescent; head liguliform, solitary, erect, the peduncle with scales, not subtended by bractlets; involucre campanulate; phyllaries up to 50, in 2–5 series, unequal, glabrous or pubescent; receptacle flat, epaleate, pitted, glabrous; ligules numerous per head, bisexual, fertile, yellow; cypselae fusiform, beakless, 10-ribbed, glabrous or scabrous; pappus of numerous white, smooth or barbellate bristles and sometimes with scales.

This genus comprises 4 species, native to North America.

Our species in the past has been placed in *Agoseris* or *Microseris*. It differs from *Agoseris* in having entire leaves and from *Microseris* in having erect flowering heads and white pappus bristles.

Only the following species occurs in Illinois.

1. **Nothocalais cuspidata** (Pursh) Greene, Bull. Cal. Acad. Sci. 2:55. 1886.
Troximum cuspidatum Pursh, Fl. Am. Sept. 2:742. 1813.
Agoseris cuspidata (Pursh) Steud. Nomencl. Bot., ed. 2, 1:37. 1840.
Microseris cuspidata (Pursh) Sch.-Bip. Jahr. Poll. 22–24:309. 1866.

Perennial herbs with a taproot; latex present; stem solitary, erect, unbranched, scapose, without scales, up to 35 cm tall, glabrous; leaves all basal, linear-lanceolate, acuminate at the apex, tapering to the sessile base, to 30 cm long, to 2.5 cm wide, conduplicate, entire, with tomentose and ciliolate margins; head liguliform, solitary, erect, to 4 cm across, the peduncle without scales, not subtended by bractlets; involucre campanulate; phyllaries up to 30 (–35), in 2–5 series, unequal, lanceolate, acuminate at the apex, usually glabrous; receptacle flat, epaleate, pitted, glabrous; ligules numerous per head, 15–25 mm long, bisexual, fertile, yellow; cypselae fusiform, without a beak, brown, 7–10 mm long, 10-ribbed, usually glabrous; pappus of numerous white unequal capillary bristles and several setiform scales.

Common Name: Prairie dandelion.
Habitat: Dry prairies.
Range: Wisconsin to Saskatchewan, south to New Mexico, Texas, and Arkansas.
Illinois Distribution: Very rare in the northern half of Illinois.

This plant looks superficially like *Taraxacum officinalis*, but the entire leaves immediately lead one to *Nothocalais cuspidata*. The prairie habitat is also distinctive.

This handsome species of prairies is one of the rarest plants in Illinois. George Vasey first found it in McHenry County in 1858.

Nothocalais cuspidata flowers from April to June.

118. **Krigia** Schreb.—Dwarf Dandelion

Annual or perennial herbs with rhizomes, tubers, or fibrous roots; latex present; stems erect, branched or unbranched, sometimes scapose, glabrous or hispidulous, often glaucous; leaves basal or with a few alternate cauline leaves, usually petiolate, entire to denticulate to pinnatifid, mostly glabrous, sometimes glaucous; heads liguliform, solitary or several in a corymb, the peduncle without scales, not subtended by bractlets; involucre campanulate; phyllaries 9–18, in 1 or 2 series, equal, glabrous; receptacle flat, epaleate, pitted, glabrous; ligules usually 25–50, erose at the summit, bisexual, fertile, yellow or orange; cypselae turbinate or columnar or oblongoid, brown or red-brown, not beaked, 10- to 20-ribbed, glabrous; pappus in 2 series, the outer of chafflike scales, the inner of capillary bristles.

I recognize 6 species in this genus, all native to North America. I am recognizing *Serinia* as a separate genus because of the absence of pappus, the fewer ribs on the cypselae, and the fewer phyllaries.

Three species of *Krigia* occur in Illinois.

1. Plants scapose at flowering time, or at least with most of the leaves at or near the base of the plant.
  2. Flowering head solitary; involucre 9–14 mm high; inner pappus with 15–40 bristles . . . . 1. *K. dandelion*
  2. Heads two or more; involucre 4–7 mm high; inner pappus with 5 bristles . . . . 3. *K. virginica*
1. Plants leafy to the summit at flowering time . . . . 2. *K. biflora*

1. **Krigia dandelion** (L.) Nutt. Gen. N. Am. Pl. 2:127. 1818.

Leontodon dandelion L. Sp. Pl. 2:798. 1753.
Cynthia dandelion (L.) DC. Prodr. 7:89. 1838.
Adopogon dandelion (L.) Kuntze, Rev. Gen. Pl. 304. 1891.

Perennial herbs from slender rhizomes and spherical tubers; latex present; stems scapose, erect, unbranched, to 45 cm tall, glabrous, often glaucous, without scales; leaves all basal, spatulate to linear-lanceolate, acute to acuminate at the apex, tapering to the winged petiole, to 20 cm long, to 7 cm wide, entire or denticulate or pinnatifid, glabrous, often glaucous; head liguliform, solitary, to 2 cm across; involucre narrowly campanulate, 9–14 mm high; phyllaries about 15, in 1 or 2 series, equal, linear-lanceolate, glabrous, reflexed at maturity; receptacle flat, epaleate, pitted, glabrous; ligules up to 30 (–34), to 2.5 cm long, bisexual, fertile, yellow or orange; cypselae oblongoid to columnar, red-brown, 2.0–2.5 mm long, with 10–20 ribs; pappus in 2 series, the outer of 10–15 scales up to 1 mm long, the inner of 15–40 white barbellulate capillary bristles 5–8 mm long.

Common Name: Dwarf dandelion; potato dandelion.
Habitat: Open woods; exposed blufftops.
Range: West Virginia to Kansas, south to Texas and Florida.
Illinois Distribution: Occasional in the southern one-third of Illinois.

This species is distinguished by its solitary flowering head at the summit of leafless stems, the presence of spherical tubers, the narrow involucre 9–14 mm high, and the inner pappus of 15–40 barbellulate capillary bristles. The stems are frequently glaucous.

This species and *Krigia biflora* differ from *K. virginica* in their perennial habit, their several-ribbed cypselae, and the 15–40 inner bristles of the pappus. These characters have led some botanists in the past to place them in the genus *Cynthia*, a decision I think may be warranted.

Krigia dandelion flowers from April to June.

2. **Krigia biflora** (Walt.) S. F. Blake, Rhodora 17:135. 1915.
Tragopogon virginicum L. Sp. Pl. 2:789. 1753, *non Hyloseris virginica* L. (1753).
Hyloseris biflora Walt. Fl. Carol. 194. 1788.
Hyloseris amplexicaulis Michx. Fl. Bor. Am. 2:87. 1803.
Krigia amplexicaulis (Michx.) Nutt. Gen. 2:127. 1818.
Cynthia virginica (L.) D. Don, Edinb. Phil. Journ. 12:309. 1829.
Adopogon virginicum (L.) Kuntze, Rev. Gen. Pl. 304. 1891.
Krigia biflora (Walt.) S. F. Blake f. *glandulifera* Fern. Rhodora 37:337. 1935.

Perennial herbs from a thickened caudex and fibrous roots; latex present; stems erect, branched above, to 75 cm tall, glabrous or nearly so or sometimes minutely glandular-pubescent, glaucous; leaves basal and cauline, the basal lanceolate, obtuse to acute at the apex, tapering to the winged petiole, to 20 cm long, to 5 cm wide, entire, denticulate, or runcinate, glabrous, glaucous, the cauline 1–4 in number, lanceolate, acute at the apex, auriculate-clasping at the base, usually entire, glabrous, glaucous, smaller than the basal leaves; heads liguliform, 1 to several in a corymb, up to 16 mm across, the peduncles glabrous or rarely glandular; involucre campanulate, 7–11 mm high; phyllaries up to 18, in 1–2 series, equal, lanceolate, glabrous, reflexed at maturity; receptacle flat, epaleate, pitted, glabrous; ligules up to 60 per head, 15–25 mm long, bisexual, fertile, yellow or orange; cypselae oblongoid or columnar, red-brown, 2.0–2.5 mm long, with 10–15 ribs; pappus in 2 series, the outer of 10–15 scales up to 0.5 mm long, the inner of numerous white capillary bristles 4.5–5.5 mm long.

Common Name: False dandelion.
Habitat: Open woods, prairies, savannas.
Range: New York and Ontario to Manitoba, south to Arizona, New Mexico, Mississippi, and Georgia.
Illinois Distribution: Common throughout Illinois.

This species differs from *K. dandelion* in having several flowering heads and a few cauline leaves. Very uncommon plants that have glandular peduncles have been called f. *glandulifera*.

As seen from the nomenclature above, this species has a long history of name changes. Placing it and *K. dandelion* in the genus *Cynthia* is tempting.

Krigia biflora flowers from May to September.

3. **Krigia virginica** (L.) Willd. Sp. Pl. 3:1618. 1803.
Hyloseris virginica L. Sp. Pl. 2:809. 1753.
Hyoseris caroliniana Walt. Fl. Carol. 194. 1788.
Krigia caroliniana (Walt.) Nutt. Gen. 2:126. 1828.
Adopogon caroliniana (Walt.) Britt. Mem. Torrey Club 5:346. 1894.

Annual herbs with a taproot; latex present; stems erect, unbranched, scapose, becoming leafy after anthesis, to 30 cm tall, glabrous but usually stipitate-glandular at the summit, usually glaucous; basal leaves spatulate to lanceolate, obtuse to acute at the apex, tapering to the winged petiole, to 15 cm long, to 5 cm wide, entire or denticulate or pinnatifid, usually glabrous, the cauline leaves, developing after anthesis, alternate, oblanceolate, to 15 cm long, entire or denticulate or pinnatifid, glabrous or sometimes pubescent; head liguliform, solitary, to 15 mm across, the peduncle usually glabrous; involucre campanulate, 4–7 mm high; phyllaries up to 18, in 1–2 series, lanceolate, glabrous, equal, reflexed at maturity; receptacle flat, epaleate, pitted, glabrous; ligules up to 35 per head, bisexual, fertile, yellow; cypselae turbinate, red-brown, 1.5–2.5 mm long, truncate at the apex, 5-angled, with 15–20 ribs; pappus in series, the outer of 5 scales up to 1 mm long, the inner of 5 white capillary bristles 4–6 mm long.

Common Name: Dwarf dandelion.
Habitat: Sandy soils, sand dunes.
Range: Maine to Minnesota, south to Texas and Florida; British Columbia.
Illinois Distribution: Occasional in the northern and western counties, rare
    elsewhere.

This *Krigia* is distinguished by the 5 inner pappus capillary bristles, the 5-angled cypselae, and the stems that become leafy after anthesis.
    Krigia virginica flowers from April to August.

119. **Serinia** Raf.—Dwarf Dandelion

Annual herbs with a taproot; latex present; stems erect, branched, glabrous or glandular-pubescent; leaves basal and cauline, the basal entire or pinnatifid, with unwinged petioles, the cauline alternate or sometimes subopposite, sessile or slightly clasping, entire; head liguliform, solitary on each branch, the peduncle leafless, not subtended by bractlets; involucre campanulate; phyllaries usually 8, in 1–2 series, equal, concave during fruiting; receptacle flat, epaleate, pitted, glabrous; ligules up to 35 per head, bisexual, fertile, yellow or orange; cypselae obovoid, red-brown, truncate at the apex, usually 10-ribbed; pappus absent.

Serinia is often included within *Krigia* but differs in the absence of a pappus and fewer phyllaries per flowering head that are not reflexed in fruit. *Serinia* is also similar to *Lapsana*, but the peduncles in *Lapsana* have bractlets at the base.
    This genus comprises 3 species, all native to the southern United States. Only the following occurs in Illinois.

1. **Serinia cespitosa** Raf. Fl. Ludov. 149. 1847.
Krigia oppositifolia Raf. Fl. Ludov. 57. 1817, misapplied.
Adopogon humilis Ell. Bot. S. C. & Ga. 2:267. 1824.
Serinia oppositifolia (Raf.) Kuntze, Rev. Gen. Pl. 364. 1891.
Krigia cespitosa (Raf.) K. L. Chambers, Journ. Arn. Arb. 54:52. 1973.

Annual herbs with a taproot; latex present; stems erect, slender, branched from the base, to 25 cm tall, glabrous or glandular-pubescent; leaves basal and cauline, the basal oblanceolate to spatulate, acute at the apex, tapering to the unwinged petiole, to 15 cm long, to 3 cm wide, entire or with an occasional tooth, glabrous or occasionally glandular-pubescent, the cauline alternate or subopposite, entire, sessile or subclasping, smaller than the basal leaves; head liguliform, solitary on each branch, to 15 mm across, the peduncle without bractlets at the base; involucre campanulate, 2–5 mm high; phyllaries usually 8, in 1–2 series, lanceolate, glabrous, not reflexed in fruit; receptacle flat, epaleate, pitted, glabrous; ligules up to 35 per head, to 10 mm long, bisexual, fertile, yellow or orange; cypselae obovoid, red-brown, 1.5–1.7 mm long, truncate at the apex, usually 10-ribbed, glabrous; pappus absent.

Common Name: Dwarf dandelion.
Habitat: Moist ground, sandy soil.
Range: Virginia to Illinois to Nebraska, south to Texas and Florida.
Illinois Distribution: Occasional in the southern half of Illinois.

This small species differs from species of *Krigia* in the absence of pappus and the 8 phyllaries per flowering head that are erect during fruiting. Many botanists keep this species in *Krigia*.

Serinia cespitosa flowers from May to July.

120. **Pyrrhopappus** DC.—False Dandelion

Annual or perennial herbs with a taproot or rhizomes; latex present; stems erect, branched or unbranched, sometimes scapose; leaves basal and/or cauline, the basal petiolate, the cauline sessile, entire to denticulate to pinnatifid; heads liguliform, 1 to few in a corymb, the peduncles subtended by bractlets; involucre cylindric or campanulate; phyllaries up to 20, in 2 series, equal, scarious on the margins, reflexed during fruiting; receptacle convex, epaleate, pitted, glabrous; ligules numerous per head, bisexual, fertile, yellow (in Illinois) or whitish; cypselae obovoid to fusiform, with a long beak, red-brown or stramineous, 5-ribbed; pappus in 2 rows, the outer of a few short whitish scales, the inner of numerous reddish capillary barbellulate bristles.

Pyrrhopappus consists of 4 or 5 North American species. It is distinguished by its long-beaked cypselae and its unequal pappus.

Only the following species occurs in Illinois.

1. **Pyrrhopappus carolinianus** (Walt.) DC. Prodr. 7:144. 1838.
Leontodon carolinianum Walt. Fl. Carol. 192. 1788.
Sitialis caroliniana (Walt.) Raf. New Fl. N. Am. 4:85. 1836.

Annual or perennial herbs from a taproot; latex present; stems erect, branched from near the base, to 90 cm tall, glabrous or nearly so; leaves basal and cauline, the basal several, oblong to lanceolate, acute to acuminate at the apex, tapering to the usually winged petiole, to 20 cm long, to 3 cm wide, glabrous, entire or denticulate or pinnatifid, the cauline few, alternate, lanceolate, acute to acuminate at the apex, tapering to the sessile and sometimes clasping base, smaller than the basal leaves, glabrous, entire, denticulate or pinnatifid; heads liguliform, 1 to several in a corymb, to 3 cm across, the peduncles subtended by up to 16 bractlets up to 12 mm long; involucre usually campanulate, 16–25 mm high; phyllaries up to 20, in 2 series, linear, equal, with scarious margins, reflexed in fruit; receptacle convex, epaleate, pitted, glabrous; ligules numerous per head, to 2 cm long, bisexual, fertile, pale yellow; cypselae fusiform, long-beaked, red-brown, 5-ribbed, 4–6 mm long, the beak 8–10 mm long; pappus in 2 rows, the outer of a few short whitish scales, the inner of numerous reddish capillary barbellulate bristles.

Common Name: False dandelion.
Habitat: Dry woods, prairies, roadsides.
Range: Pennsylvania to Nebraska, south to Texas and Florida.
Illinois Distribution: Occasional to common in the southern two-thirds of the state, absent elsewhere.

This species differs from all other members of the Cichorieae in Illinois in its long-beaked cypselae and its reddish inner capillary bristles of the pappus.

    Pyrrhopappus carolinianus flowers during May and June.

Excluded Species

Agoseris glauca (Pursh) Raf. Kibbe in 1952 called *Nothocalais cuspidata* by this binomial.

Anaphalis margaritacea (L.) Gray. Several Illinois botanists used this binomial for *Pseudognaphalium obtusifolium* Hilliard & R. L. Burtt., but that is a different species.

Antennaria dioica (L.) Gaertn. This binomial was used by Mead in 1846 and Lapham in 1857 for what is now *Antennaria howellii* Greene ssp. *neodioica* (Greene) R. J. Bayer.

Antennaria margaritacea (L.) R. Br. Schneck in 1876 and Patterson in 1876 erroneously used this binomial for what is now *Pseudognaphalium obtusifolium* Hilliard & R. L. Burtt.

Cirsium lanceolatum Hill. Illinois botanists up until 1939 used this binomial for *C. vulgare* (Savi) Tenore, but Hill's plant is a different species.

Cirsium pumilum (Nutt.) Spreng. Illinois botanists during the first half of the twentieth century used this binomial for *C. hillii* (Canby) Fern.

Cirsium virginianum (L.) Michx. Mead in 1846 and Lapham in 1857 used this binomial for *C. hillii* (Canby) Fern.

Cnicus pumilus (Nutt.) Torr. This binomial was used by Brendel in 1887 and Huett in 1897 for *Cirsium hillii* (Canby) Fern.

Cnicus virginianus (L.) Pursh. Short in 1845 used this binomial for *Cirsium hillii* (Canby) Fern.

Eupatorium aromaticum L. This binomial refers to a plant in the southeastern United States. Short in 1845, Patterson in 1876, and Brendel in 1887 used this name for *Ageratina altissima* (L.) R. M. King & H. Robins.

Hieracium venosum L. was used erroneously by McDougall in 1936 for *Krigia biflora* (L.) S. F. Blake.

Lactuca virosa L. Gates in 1923 called *L. serriola* L. by this name. *Lactuca virosa* is a different species.

Liatris graminifolia (Willd.) Pursh. Ries in 1939 incorrectly used this name for *L. pycnostachya* Michx.

Liatris ligulistylis (Nels.) Schum. This binomial has been applied, I believe erroneously, to *L. scariosa* var. *nieuwlandii*. I believe *L. ligulistylis* refers to a different plant.

Nabalus virgatus (Michx.) DC. Lapham in 1857 used this binomial for *N. latissimus* (L.) Hook.

Pluchea foetida (L.) DC. Illinois botanists up until 1888 used this binomial for *P. camphorata* (L.) DC.

Prenanthes serpentaria Pursh. Short in 1845 and Mead in 1846 misapplied this binomial to *Nabalus albus* (L.) Hook.

Prenanthes virgata Michx. Short in 1845 erroneously used this binomial for *Nabalus altissimus* (L.) Hook.

Vernonia corymbosa Schw. Short in 1845 used this binomial for *V. fasciculata* Michx., but *V. corymbosa* is a different species.

Vernonia noveboracensis (L.) Michx. Illinois botanists up until 1937 mistakenly used this binomial for *V. missurica* Raf.

Glossary

acuminate. Gradually tapering to a long point.

acute. Sharply tapering to a point.

annual. A plant that lives for only one growing season.

anthesis. Flowering time.

apiculate. Abruptly short-pointed at the tip.

appressed. Lying flat against the surface.

aristate. Bearing an awn.

array. An arrangement of flowering heads in an inflorescence in the Asteraceae.

attenuate. Gradually becoming narrowed.

auriculate. Bearing an earlock process.

awn. A bristle usually terminating a structure.

axil. The junction of two structures.

axillary. Borne in an axil.

barbellate. Bearing barbs or setae with downward-pointing prickles.

barbellulate. Bearing small barbs or small setae with downward-pointing prickles.

beaked. Having a short point at the tip.

biennial. A plant that completes its life cycle in two years and then perishes.

bisexual. Said of a flower bearing both staminate and pistillate parts.

bract. An accessory structure on the peduncle bearing flowering heads.

bracteate. Bearing one or more bracts.

bracteolate. Bearing one or more bracteoles.

bracteole. A secondary bract.

bractlet. A small bract.

bristle. A stiff hair or hairlike growth; a seta.

caducous. Falling away early.

calyculus. A bractlet at the base of a peduncle.

campanulate. Bell-shaped.

canescent. Grayish-hairy.

capillary. Threadlike.

capitate. Forming a head.

capitulum. A flowering head in the Asteraceae.

caudex. The woody base of a perennial plant.

cauline. Belonging to a stem.

cespitose. Growing in tufts.

chaff. A scale or group of scales.

chartaceous. Growing in tufts.

cilia. Marginal hairs.

ciliate. Bearing cilia.

ciliolate. Bearing small cilia.

clasping. Said of a leaf whose base wraps partway around the stem.

columnar. Shaped like a column or cylindrical upright structure.

compressed. Flattened.

concave. Curved on the inner surface; opposed to convex.

conduplicate. Folded lengthwise.

convex. Curved on the outer surface; opposed to concave.

cordate. Heart-shaped.

coriaceous. Leathery.

corolla. That part of a flower composed of petals.

corymb. A type of inflorescence where the pedicellate flowers are arranged along an elongated axis but with the flowers all attaining about the same height.

corymbiform. Shaped like a corymb.

crenate. With round teeth.

cuspidate. Terminating in a very short point.

cyme. A type of broad and flattened inflorescence in which the central flowers bloom first.

cymose. Bearing a cyme.

cypsela. The fruit in the Asteraceae.

deciduous. Falling away.

decumbent. Lying flat but ascending at the tip.

decurrent. Adnate to the petiole or stem and then extending beyond the point of attachment.

decussate. Arranged along the stem in pairs, each pair at right angles to the pair next above or below.

dentate. With sharp teeth, the tips of which project outward.

denticulate. With small, sharp teeth, the tips of which project outward.

diffuse. Loosely spreading.

disc. The group of flowers that make up the center of a flowering head in the Asteraceae.

disciform. A flowering head with only disc flowers but with peripheral flowers that have filiform corolla that are usually only pistillate.

discoid. Bearing only disc flowers.

eciliate. Without cilia.

eglandular. Without glands.

ellipsoid. Referring to a solid object that is broadest at the middle, gradually tapering to both ends.

elliptic. Broadest at the middle, gradually tapering to both ends.

entire. Without any projections along the edge.

epaleate. Without palea or outgrowths from the receptacle in the Asteraceae.

epunctate. Without dots.

erose. With an irregularly notched tip.

fascicle. Cluster.

fibrous. Referring to roots borne in tufts.

filiform. Threadlike.

flexuous. Zigzag.

fusiform. Spindle-shaped.

glabrate. Becoming smooth.

glabrous. Without pubescence or hairs.

gland. An enlarged, usually spherical body functioning as a secretory organ.

glandular. Bearing glands.

glaucous. With a whitish covering that may be rubbed off.

globose. Round; globular.

glomerule. A small, compact cluster.

glutinous. Covered with a sticky secretion.

hirsute. Bearing stiff hairs.

hirtellous. Finely hirsute.

hispid. Bearing rigid hairs.

hyaline. Transparent.

inferior. Referring to the position of the ovary when it is surrounded by the adnate portion of the floral tube or is embedded in the receptacle.

inflorescence. A cluster of flowers.

involucre. The collection of phyllaries around a flower cluster in the Asteraceae.

involute. Rolled inward.

keel. A ridgelike process.

laciniate. Divided into narrow, pointed divisions.

lamina. A blade.

lanceolate. Lance-shaped; broadest near the base, gradually tapering to the narrower apex.

lanceoloid. Referring to a solid object that is broadest near the base, gradually tapering to the narrower apex.

latex. Milky sap.

leaflet. An individual unit of a compound leaf.

ligulate. Flowers with ligules.

ligule. A flat, narrow petal-like structure of a flower in the Asteraceae; a ray.

linear. Elongated and narrow in width throughout.

lustrous. Shiny.

mucro. A short, abrupt tip.

mucronate. Possessing a short, abrupt tip.

mucronulate. Possessing a very short, abrupt tip.

obconic. Reverse cone-shaped.

oblanceolate. Reverse lance-shaped; broadest at apex, gradually tapering to narrow base.

oblong. Broadest at the middle and tapering to both ends, but broader than elliptic.

oblongoid. Referring to a solid object that is broadest at the middle and tapering to both ends.

obovate. Broadly rounded at the apex, becoming narrowed below.

obovoid. Referring to a solid object that is broadly rounded at the apex, becoming narrowed below.

obtuse. Rounded at the apex.

orbicular. Round.

oval. Broadly elliptic.

ovate. Broadly rounded at base, becoming narrowed above; broader than lanceolate.

ovoid. Referring to a solid object that is broadly rounded at base, becoming narrowed above.

palea. An outgrowth from the receptacle in the Asteraceae.

paleate. Bearing paleae.

palmate. Divided radiately, like the fingers from a hand.

panicle. A type of inflorescence composed of several racemes.

paniculate. Having the flowers in a panicle.

pannose. Having the texture of felt.

pappus. Various kinds of structures attached to the cypsela in the Asteraceae.

pedicel. The stalk of a flower.

pedicellate. Said of a flower that has a pedicel.

peduncle. The stalk of an inflorescence.

perennial. Living more than two years.

perfect. Said of a flower that has both stamens and pistils.

petiolate. Having a petiole.

petiole. The stalk of a leaf.

phyllary. A bract subtending a flowering head in the Asteraceae.

pilose. Bearing soft hairs.

pinna. A primary division of a compound blade.

pinnate. Divided once into distinct segments.

pinnatifid. Said of a simple leaf or leaf part that is cleft or lobed only partway to its axis.

pinnatisect. Divided in a pinnate manner.

pistillate. Bearing pistils but not stamens.

plumose. Bearing fine hairs, like the plume of a feather.

prostrate. Lying flat.

puberulent. Bearing minute hairs.

pubescent. Bearing some kind of hairs.

punctate. Dotted.

pustulate. Having small, pimplelike swellings.

pyramidal. Shaped like a pyramid.

raceme. A type of inflorescence where pedicellate flowers are arranged along an elongated axis.

radiate. Bearing ray flowers in the Asteraceae.

ray. A flat flower in the Asteraceae; a ligule.

receptacle. A structure to which all flowers are attached in the capitulum in the Asteraceae.

reflexed. Turned downward.

resinous. Producing a sticky secretion, or resin.

reticulate. Resembling a network.

rhizomatous. Bearing rhizomes.

rhizome. An underground horizontal stem bearing nodes, buds, and roots.

rosette. A cluster of leaves in a circular arrangement at the base of a plant.

rugose. Wrinkled.

rugulose. With small wrinkles.

runcinate. Pinnately cleft, with the lobes curved backward.

sagittate. Shaped like an arrowhead.

scabrellous. Slightly rough to the touch.

scabrous. Rough to the touch.

scarious. Thin and membranous.

secund. Borne on one side.

septate. With dividing walls.

sericeous. Silky; bearing soft, appressed hairs.

serrate. With teeth that project forward.

serrulate. With very small teeth that project forward.

sessile. Without a stalk.

seta. Bristle.

setaceous. Bearing bristles or setae.

setose. Bearing setae.

sordid. Dirty, in tint.

spatulate. Oblong but with the basal end elongated.

spicate. Bearing a spike.

spike. A type of inflorescence where sessile flowers are arranged along an elongated axis.

spinescent. Becoming spiny.

spinose. Bearing spines.

spinulose. Bearing small spines.

squarrose. With tips turned back or spreading.

staminate. Bearing stamens but not pistils.

stipe. A stalk.

stipitate. Bearing a stalk.

stolon. A slender, horizontal stem on the surface of the ground.

stoloniferous. Bearing stolons.

stramineous. Straw-colored.

striate. Marked with grooves.

strigillose. Bearing short, appressed, straight hairs.

strigose. Bearing appressed, straight hairs.

subacute. Nearly tapering to a short point.

subcordate. Nearly cordate.

suborbicular. Nearly round.

subulate. With a very short, narrow point.

succulent. Fleshy.

terete. Round in cross section.

ternate. Divided three times.

thyrse. A mixed inflorescence containing panicles and cymes.

thyrsoid. Bearing a thyrse.

trigonous. Triangular in cross section.

truncate. Abruptly cut across.

turbinate. Top-shaped; shaped like a turban.

undulate. Wavy.

urceolate. Urn-shaped.

villous. Bearing long, soft, slender, unmatted hairs.

viscid. Sticky.

whorl. An arrangement of three or more structures at a point.

Literature Cited

Brendel, F. 1887. *Flora Peoriana*. Peoria, Ill.: published by the author.

Flora of North America. Vol. 20, *Magnoliophyta: Asteridae (in part): Asteraceae, Part 2*. 2006. Oxford: Oxford University Press.

Gaiser, L. O. 1946. The genus *Liatris*. *Rhodora* 48:165–83; 216–63; 273–326; 331–82; 393–412.

Gates, F. C. 1923. Flora of Cass County, Illinois. *Transactions of the Illinois Academy of Sciences* 15:165–70.

Huett, J. W. 1897. Essay toward a natural history of LaSalle County, Illinois. *Flora LaSallensis*, part 1. LaSalle, Ill.: published by the author.

Kibbe, A. L. 1952. *A botanical study and survey of a typical mid-western county (Hancock County, Illinois)*. Carthage, Ill.: published by the author.

Lapham, I. A. 1857. Catalogue of the plants of the state of Illinois. *Transactions of the Illinois State Agricultural Society* 2:492–550.

McDougall, W. B. 1936. *Field book of Illinois wild flowers*. 406 pp. Urbana: Illinois Natural History Survey.

Mead, S. B. 1846. Catalogue of plants growing spontaneously in the state of Illinois, the principal part near Augusta, Hancock County. *Prairie Farmer* 6:35–36; 66; 93; 119–22.

Patterson, H. N. 1876. *Catalogue of the phaenogamous and vascular cryptogamous plants of Illinois*. 54 pp. Oquawka, Ill.: published by the author.

Ries, D. T. 1939. Additions to the flora of Starved Rock. *Transactions of the Illinois Academy of Sciences* 32:89–90.

Schneck, J. 1876. Catalogue of the flora of the Wabash Valley. *Annual Report of the Geological Survey of Indiana* 7:504–79.

Short, C. W. 1845. Observations on the botany of Illinois. *Western Journal of Medicine and Surgery* 3:185–98.

Index of Scientific Names

Names in roman type are accepted names, while those in italics are synonyms and are not considered valid.

Index of Common Names

Robert H. Mohlenbrock taught botany at Southern Illinois University Carbondale for thirty-four years. Since his retirement in 1990, he has served as senior scientist for Biotic Consultants, teaching wetland identification classes around the country. Among his more than sixty books are *Vascular Flora of Illinois*, *Where Have All the Wildflowers Gone?*, and *Field Guide to the U.S. National Forests*.